noel Chado
With compl

£3.50

POSITIONAL ASTRONOMY
AND
ASTRO-NAVIGATION MADE EASY

A new approach using the Pocket Calculator

Figures 2.3–2.6 are reproduced by
permission of George Philip & Son Limited,
12–14 Long Acre, London WC2E 9LP

POSITIONAL ASTRONOMY AND ASTRO-NAVIGATION MADE EASY

A new approach using the Pocket Calculator

H. R. MILLS, OBE, MSc, DipEd, CEng, MIEE, FRAS

Stanley Thornes (Publishers) Ltd

Text and photographs © H. R. Mills 1978
Line illustrations © Stanley Thornes (Publishers) Ltd.

All rights reserved. No part of this publication may be reproduced, stored in a retrieval system or transmitted in any form or by any means, electronic, mechanical, photocopying, recording or otherwise without the prior written permission of the copyright holder and publisher.

First published in 1978 by:
Stanley Thornes (Publishers) Ltd.,
Educa House, Kingsditch, Cheltenham,
Glos. GL51 9PL, England

ISBN 085950 062 4

Typeset by Gloucester Typesetting Co. Ltd.
Reproduced and printed by Gloster Graphics Ltd.
Bound by The Pitman Press, Bath.

Foreword by Dr. D. McNally

Positional Astronomy is often considered to be a 'difficult' subject. There are several well found reasons for this view. The subject is precise in concept. The need for precision of concept often obscures a straightforward underlying idea. There is a need for trigonometric formula associated with the geometry of the sphere. Many give up in the face of formula complicated in appearance and tedious to evaluate even if simple to derive. Add to this the fact that precise astrometry demanded painstakingly laborious work with a measuring engine and it is little wonder Positional Astronomy got a poor reputation. That reputation was compounded since astrometry also came to be regarded as 'old fashioned' by comparison with astrophysics.

Times change and modern instrumentation and computational expertise are taking the labour out of astrometric measurements. It is now the situation that astrometry appears poised to enter a period of unprecedented development. Once again, astrometry is being equipped to tackle the most fundamental problems of astronomical science.

Mr. Mills' book, therefore, comes at an opportune time. He shows that with a modern hand held calculator and some simple apparatus, the basic methods of positional astronomy are easily accessible. The high precision that can cloud an otherwise straightforward concept has been omitted. The book makes no claims to satisfy the finesse demanded by modern astronometry but supplies a straightforward account of the calculations required so that navigators can navigate with sufficient precision to avoid difficulty and astronomers can direct telescopes to acquire specific stars with ease.

This book is particularly apt for school use. The simple apparatus is cheap to build and reliable in use (to the accuracy intended). The concepts used are adequate for the job. The tedious mathematics is removed by use of a simple calculator. Therefore, any senior student who understands a modicum of trigonometry can set about extended calculations. It is a valuable thing to have practice in extended calculation and there are few situations apart from Positional Astronomy which offer realistic practice at this level. One notes the inability of many science students at University to tackle anything but the simplest numerical problems — indeed many students seem unwilling to think at all in numerical terms.

It is my hope that Mr. Mills will contribute to a revival of interest in Positional Astronomy in a new generation. I hope that some may be inspired to take Positional Astronomy to a professional level and ultimately make a worthwhile contribution to understanding the structure of the Universe through their use of this important astronomical technique.

Dr. D. McNally

University of London Observatory   Spring 1978

Acknowledgements

I would like to express my thanks to the Salisbury College of Technology for the use of its London linked computer facilities through the cooperation and help of Mr. R. Wood of the Engineering Staff; to the College of Art for several photographs by Mr. J. Lamb; to Mr. E. Butt for his help with photographic work; to Mr. Brian Foster of the Salisbury Astronomical Society for the line drawings; to Mrs. Mary Smith for working wonders with the typewriter on my illegible script; to many members of the British Astronomical Association who have always been helpful with encouragement, criticism and friendly advice; and to my wife for her help and forbearance consistently given throughout the work of compiling the material and making the models. I would like to record my appreciation of the friendly cooperation and guidance received from the publishers.

Finally, I am particularly grateful to Dr. D. McNally, Chairman of the Education Committee of the Royal Astronomical Society, and formerly President of the Teaching Commission of the International Astronomical Unions, for his initial interest and continued guidance without which I would not have had the temerity to write about some of the fun I have had in compiling this book.

A note on the career of H. R. Mills

Mr. H. R. Mills, OBE, MSc, DipEd, CEng, MIEE, FRAS, formerly held the posts:

Principal of the Cochin State College in South India, under the University of Madras.

Instructor Lt. Commander in the Royal Indian Navy (1939–46).

Director of the Science and Engineering Department of the British Council.

Scientific Adviser to the British Government for the Colombo Plan and technical aid in S. and S.E. Asia.

UNESCO adviser on Vocational Education and Training.

He has spent over 30 years in the tropics where ideal observing conditions encouraged his interest in astronomy.

Contents

Chapter 1 INTRODUCTION   1

Object of book; the calculator explosion; new possibilities for projects for schools and colleges; astronomy in educational programmes; astronomy as a basis for an integrated science programme; a note on calculators; the many branches of astronomy.

Chapter 2 THE CELESTIAL SPHERE   9

Understanding the celestial sphere; astronomical terms and fundamental concepts; astronomical ephemerides; the star globe; the planisphere; a nocturnal to tell the time from the stars; the Greenwich hour angle or GHA; sidereal time; sidereal hour angle and the use of the *Nautical Almanac*; reducing sidereal time to mean time; finding the Greenwich hour angle; a home made meridian line; spherical trigonometrical formulae regarded as calculator programmes; importance of understanding principles.

Chapter 3 SPHERICAL TRIANGLES AND EARLY DEVICES FOR THEIR SOLUTION   29

The story of spherical triangles; Pythagoras; stereoscopic projection; a simple shadow astrolabe; a protractor astrolabe; a cross staff astrolabe; an alt–azimuth star finder; model sextant; fundamental formulae; the basic astro-navigation formula; a summary of useful relations used in positional astronomy and astro-navigation; sight reduction tables; finding hour angles from ephemerides; times of rising and setting of celestial bodies; amplitudes or bearings of the sun when rising or setting; the prime vertical; the four part formula; a diagram showing altitude as the head of a family of mathematical relations; megalithic astronomy; the four main coordinate systems used in positional astronomy; model of an alt–azimuth device; setting circles; angles of stars at their risings and settings; effect of refraction; how much sunshine can be expected at a particular latitude; the hour angle of bodies at rising and setting for various declinations; sunrise, sunset and twilight.

Chapter 4 ASTRO-NAVIGATION WITH A CALCULATOR   77
Great circle sailing; an example of the calculator working; position line in 11 steps; examples; plotting the position lines; precautions; dip; atmospheric refraction; index error of sextant; semi diameter of Sun or Moon; accurate timing of observation; the Moon's parallax; practice with a sextant – on land; an artificial horizon; a calculator cannot improve observational accuracy; an automatic digital sextant; Polaris and the calculator; finding the observer's latitude; traverse and departure; rhumb line and course using a calculator; Mercator charts; the steady course and how it can be calculated; a suggested method of recording data for position lines; calculation of the azimuth of a body; finding the compass error; deviation of the compass and the variation; the prime vertical altitude; finding due east or west; a simple navigational puzzle.

Chapter 5 ALTITUDE AND AZIMUTH LINES   113
Star maps and the limitations of a polar type planisphere; the curves plotted on polar graph paper; the horizon curve; almucantars or altitude curves; tables for $\phi = 40°$ 51°, 53°, 55° and 57°; almucantars for $\phi = 51°$, 53°, 55°, and 57°; drawing the curves; the overlay in position; applicability of results to both northern and southern hemispheres; plotting the curves using centimetric rectangular coordinates; plotting the curves on a Mercator star map; the planisphere used as a sun compass by means of the alt–azimuth overlay graticule; the false 'watch compass'; table relating Sun's azimuth to hour angle of the Sun; construction and use of sun compasses; summary of the ways in which alt–azimuth lines can be inscribed on a graticule using polar, rectangular or Mercator coordinates.

Chapter 6 THE MEDIAEVAL ASTROLABE   147
A new approach; history; stereographic projection; the new approach to marking an astrolabe as a project; the calculated astrolabe; drawing the azimuth circles; the calculation for radii and positions of centres; practical instruction; the completed azimuth pattern; the centres and radii derived from trigonometry; the almucantars or altitude circles; altitude circles by trigonometrical formulae; the astrolabe, rete or star map; the astrolabe – using the alt–azimuth tables.

Chapter 7 PROJECTS WITH SUNDIALS AND THE
CALCULATOR   165

The measurement of time; the stick in the sand, used as a sundial; the style pointing to the pole; the three main types of sundial; the equatorial sundial; the horizontal dial; the south facing vertical dial; the vertical south facing dial – declining; the formulae necessary for the vertical dial displaced by an angle θ from the east–west curve; altitude sundials; the pillar dial; the disc dial; a simple project for a 'pocket' sundial; the Capuchin dial; the polar dial; the azimuth dial; the elliptical or analemmatic dial; a possible project concerning the analemmatic dial; declination lines – for a vertical south facing dial; declination lines for a horizontal dial.

Chapter 8 THE EQUATION OF TIME   209

Time; the mean solar day; the Earth's rotation; the Sun's apparent irregularities; the dynamical system; the problem has two parts; the eccentricity of the Earth's orbit; the obliquity of the ecliptic; the approximate formula for E; graphs showing E_1 and E_2, and E; a slightly more simplified formula; examples of various calculations; noon marks for a vertical dial; noon marks for a polar dial.

Chapter 9 PRECESSION OF THE EARTH'S AXIS   227

The pole of the ecliptic; the two systems of coordinates; the use of the star globe; changing from one system to the other; Polaris, Sirius and Crucis β; factors liable to produce errors in calculating.

Chapter 10 MISCELLANEOUS CALCULATIONS   239

How altitudes and azimuths change with hour angle; the use of the calculus; some simple relations; simple celestial mechanics applied to satellites; stellar magnitudes; limiting magnitudes.

POSTSCRIPT   249

APPENDICES   250

  I Derivation of the four part formula.

  II Useful information.

III Stellar distances.

IV Number of days since 21st September.

V Right Ascensions and declinations of some of the brightest stars.

VI Finding the date and hour when a star is on the meridan.

VII The 'summer triangle' solved as an exercise.

Bibliography. 263

Index. 265

A note on the text

In this book the following symbols and abbreviations are used:

Altitude (in general formulae)	(alt)
Altitude (as calculated from ϕ, HA, and δ)	Hc
Altitude (as observed and corrected for atmospheric refraction, dip, parallax and instrumental error)	Ho
Azimuth	Az
Declination	δ
Latitude of the observer and Altitude of the Pole	ϕ
Geographical longitude	L
The vernal equinox, or the first point of Aries	♈
The autumnal equinox	♎
Hour angle	HA
Hour angle of a star	HA*
Local hour angle	LHA
Local hour angle of a star	LHA*
Local hour angle of the Sun	LHA☉
Greenwich hour angle	GHA
Local sidereal time	LST
Sidereal hour angle of a star	SHA*
Right Ascension	RA
Universal time	UT
Zenith distance	ZD
Greenwich mean time	GMT
Sidereal time at 00.00 GMT	ST at 0^h

When the abbreviations have been used in mathematical formulae a bracket has been placed either side to give clearer separation from geometric points.

1 Introduction

1.1 Introduction

No one can fully understand and make useful calculations about the positions and motions of heavenly bodies without the ability to solve spherical triangles, but from the time of the ancient Greeks until a year or two ago the spherical triangle was for amateur astronomers a much maligned mathematical outcast, not easy to get on with, and to be avoided if possible. The history of astronomy and navigation abounds with ingenious devices and constructions designed to avoid or to tame the spherical triangle.

One of the objects of this book is to encourage a greater confidence in, and a wider use of, spherical triangles in astronomy by amateur enthusiasts, yachtsmen and upper school students, now that we have available a revolutionary aid of surprising capabilities in the form of the pocket electronic scientific calculator. This innovation has made child's play of the hitherto tedious and laborious calculations using spherical triangles and their associated formulae that have in the past bedevilled the lot of budding astronomers and navigators.

1.2

A calculator explosion has taken place,* and scientific calculators are now part of a science student's equipment and are widely advertised in all technical journals. Some teachers may raise their hands in horror over the possible dangers of students using calculators without fully understanding the mathematical processes involved. This view overlooks two basic conditions for sound learning: successful effort and immediate feedback of results. These two conditions help the learner to understand more easily and to gain confidence in meeting more advanced problems without having to waste time and effort on unrewarding calculations. In the past most of us have used log and trig tables by following rule of thumb procedures with no thought of how the tables have been compiled. Schoolboys often hate mathematics because for many the subject is inseparable from tedious arithmetical calculations and the discouraging and inhibiting effect of getting sums wrong.

1.3

Now that modern 'scientific' style electronic calculators are part of a science student's equipment and have rendered slide-rules, log tables, trig tables, haversines, and sight reduction and azimuth tables practically obsolete, an exciting change is taking place in our approach to positional astronomy, astro-navigation and to the teaching of astronomy. They are

*'The Calculator Explosion', *New Scientist*, Calculator Supplement, 13 November 1975.

Introduction

opening up many new possibilities for projects for schools and astronomical societies.

A small hand scientific calculator with trig functions, at a cost well within the means of most students, can be used to solve any spherical triangle in a matter of one or two minutes, correct to eight or nine significant figures, without paper work, and without the exacting interpolations inseparable from calculations involving navigation tables.

Fig. 1.1. A sample of one of the many scientific calculators which provide scope for new projects in astronomy for amateur astronomers, schools and colleges. It makes easy the solution of problems in positional astronomy for yachtsmen and all concerned with astro-navigation.

Positional Astronomy and Astro-Navigation Made Easy

There are dozens of different kinds and a wide range of prices. The numbers sold in the past five or six years all over the world run into hundreds of millions. For astronomical elementary calculations, the main requirements are the ordinary arithmetical functions together with trig and log functions. The only possible advice that can be given to a beginner is to follow the maker's manual of operation carefully, practise on examples given, and then on exercises of his own devising on the lines suggested in this book.

It is presumed in this book that the reader will have some knowledge of the basic principles of the celestial sphere and of elementary mathematics, such as that normally taught in the senior forms of schools.

It is hoped that students, amateur astronomers, and yachtsmen will be introduced to a little of the fun and satisfaction of using mathematical relations and concepts that have hitherto been the particular domain of professional mathematicians and astronomers who have the time and computer facilities for effecting the often laborious calculations met with in positional astronomy and astro-navigation.

1.4 Astronomy has hitherto been a neglected school or university subject partly because it is associated with expensive equipment, and partly because it is surrounded by an aura of esoteric mystery which in past centuries has been the domain of the high priests of science or religion. These reasons are not valid today. The ancient astronomers, who with commendable accuracy catalogued stars and planets, and made astrolabes, navigational instruments and ingenious sundials, had no optical instruments. However, they were extremely clever geometricians and careful observers and were able to cope with trigonometry without a full understanding of calculating by the formulae of spherical trigonometry. They were always held in high regard from the builders of lunar observatories such as Stonehenge, to Hipparchus and Copernicus.

Cumbersome and frightening formulae are now properly regarded as simple calculator programmes to be tapped out with one finger and without paper or pencil. This press button mathematics makes it possible for teachers to explain the essential subject matter without tedium, and for students to grasp and understand how the programmes are structured, the formulae arrived at, and the answers produced.

As mentioned, the instant feedback of the result is one of the great motivators in the learning process.

A perusal of the contents will indicate the scope of the topics dealt with and the examples worked out. This book also contains suggested projects that can be carried out in a school workshop, or on a drawing board using only ruler, protractor, compass, and polar or rectangular graph paper. Each project will provide in addition, exercises for students and others in the use of 'scientific' calculators.

Considerable satisfaction can be derived from working out examples to be found in astronomy books and navigation manuals of pre-calculator days. These examples showed laborious working involving trig and log tables or haversines that required about half an hour to work out. It is a matter for some excitement and not a little surprise to get the correct answer in less than a minute by pressing a few keys in the correct—logical—order (*see Bibliography*).

1.5   No mathematics teacher need be short of examples or projects in astronomy for exercises with a calculator. Astronomy can be and often is used as a basis for an integrated science programme, which in view of the present absorbing interest in space probes, Moon landings, lasers, black holes, radio and television programmes on astronomy, would be relevant to modern curriculum development.* Astronomy can be successfully introduced into a science programme only if presented and received with understanding and enjoyment. Beware of the disaster that follows learning without understanding, as exemplified by the boy who in answer to the question:

"What causes the tides?" wrote:

"Tides are caused by the rays of the Moon striking the surface of the sea at an angle of $23\frac{1}{2}°$ Fahrenheit"!

1.6   ## A Note on Calculators

It is important to understand and to become familiar with the mode of operation of your own calculator so that calculations are programmed to give correct results quickly and with a minimum of work with pen and paper.

Present day technology produces calculators that consist of four distinct assemblies.

(1)   A chip of silicon with the equivalent of thousands of transistors engraved on its surface which react like microscopic traffic signals in a complex maze of electronic paths.

*See Section 1.7.

(2) A diode which lights up numbers on a display panel.
(3) A keyboard.
(4) A robust plastic case.

Some potential users, particularly navigators, may entertain serious doubts about the ability of the calculator to stand up to rough treatment or temperature changes, but to quote from a recent article:
"The ruggedness of these miniature marvels is indeed remarkable. The 'brain' is embedded in a layer of protective plastic that makes it almost indestructible. A calculator lost on a 5000 ft mountain where summer temperatures soar above 37.8 °C and winter brings 10 feet of snow blinked instantly to life when retrieved 18 months later."*

Many of the calculator results are given without regard to the practical significance of all the nine figures displayed in the display panel. The last few figures in the final result may have to be pruned with common sense to appropriate significant figures when these are related to observational results and their known limits of accuracy.

A calculator recommended for use in the upper forms of schools and for dealing with the projects in this book, should have the following scientific functions: sin, cos, tan, and inverses, $\log_{10}$, 10^x, ln, e^x, y^x, x, $1/x$ and brackets.

It should be appreciated that even with the best pocket calculators small inaccuracies can develop during the progress of a long and involved computation arising from an accumulation of round-off or cut-off errors in the eight or nine figure display. The errors could become serious when several trigonometrical functions of very small angles are involved. A specially designed accuracy test for a pocket calculator suggested by a leading manufacturing company, (CBM) is given below as an exercise.
Evaluate $\quad \sin^{-1}(\cos^{-1}(\tan^{-1}(e^{\log_e \sqrt{\tan(\cos(\sin 29))}})^2))$

This expression clearly reduces to 29° but the full calculation can be carried out by pressing the keys in the following order of algebraic logic, 29°, sin, cos, tan, $\sqrt{x}$, ln, e^x, x^2 tan^{-1}, cos^{-1}, sin^{-1}. The result of this calculation was 29°.001. For this kind of calculation a limit of error of ± 0.05 is acceptable.

1.7 Astronomy embraces practically all branches of science, and these can be listed in three main divisions of astronomy, as follows. (The list is not complete and there are other systems of classification, but it suggests possibilities for astronomy as a central theme in curriculum development.)

*'Pocket Boom in Press Button Maths', *The Reader's Digest*, December 1976.

Introduction

The Main Branches of Astronomy

Descriptive	Gravitational	Physical

History
  The Ancient world
Megalithic sites
Calendars
  Astrolabes
Star catalogues
Positions of all celestial bodies
Instruments
  Telescopes—optical—radio
  Cameras and photography
Clocks—sidereal time
Navigation by stars
  Astrolabes
  Scientific electronic calculation
  Computers
Stellar magnitudes
  Colour
Star clusters
Galaxies
Earth science
Geography
Planets
Astronomical objects in works of art
The impact of astronomy on religious thought and philosphy
Science fiction
	Dynamics
	Kepler's Laws
	Newton's Laws
	Orbits under central forces
	Prediction of positions
	Distances
	Momenta
	Masses of celestial bodies
	Sizes
	Densities
	Cosmology
	Relativity
	Gravitational forces in stars' interiors
	Tides
	Cosmology
		Composition of stars
		  Abundance of elements
		Origin of stars and planets
		  Organic chemical molecules in space
		The origin of life
		Evolution
		Earth science
		Geology
		Age and stellar evolution
		  White Dwarfs
		  Red Giants
		Heat and energy production
		Temperature of stars
		Radioactivity
		Interstellar space
		Plasma—physics
		Nuclear chemistry
		  Fundamental particles
		Spectroscopy
		  The radiation spectrum from γ-rays to radio waves
		Quasars
		Pulsars—red shift
		Black holes
		Time measurement
		Propagation of light
		  Doppler effect
		Relativity
		Aberration
		Magnetic and electric fields in space
		The solar wind

These three branches are closely linked so that each one depends for its advance on research and discoveries in the other two.

2
The Celestial Sphere

2.1 The Celestial Sphere

An understanding of the celestial sphere is necessary before calculations can be satisfactorily carried out in astronomy. Most books on elementary astronomy, such as those mentioned in the Bibliography, deal with the arithmetical parts of positional astronomy, i.e., those parts which mathematically require only skill in adding or subtracting hours, minutes and seconds, or their angular equivalents, for calculating Right Ascensions, declinations, longitudes, or latitudes, sidereal times and hour angles. These requirements form a fair part of the syllabus for any elementary astronomy course. For example, the calculation of the hour angle of a heavenly body at a particular time and place is an exacting task for beginners, and if you are a beginner there is not much you can do with an hour angle when you have found it. You cannot, for example, find the body's altitude or its azimuth without the use of spherical triangles. This is where the scientific calculator is now an almost indispensable asset, in providing scope for practical projects in amateur astronomy, and for solving the many fascinating problems in positional astronomy.

2.2

Some readers may be unfamiliar with the astronomer's vocabulary, as many of the terms used have been handed down from ancient times, since astronomy is the oldest of sciences. It is recommended that beginners should study the first fifteen pages of *Norton's Star Atlas*, Sixteenth Edition, which contains a good glossary of astronomical terms and an account of fundamental concepts concerning the celestial sphere and time.

Many beginners find a star globe helpful. It is a valuable visual aid to understanding positional astronomy. A star globe, Fig. 2.1 is turned so that stars appear to rotate on an axis of the globe running through the globe from the North Pole point (very near the Pole Star) to the South Pole point. There does not happen to be a convenient visible star at or very near the South Pole of the celestial sphere. Stars can be given positions by the use of numbers in the same way as towns and places can be given positions by latitude and longitude. The equivalent longitude lines all run from the North Pole to the South Pole, and the latitude lines are a series of parallel belts numbered from the central belt or equator. Study this for a few minutes on a globe of the world and compare with the celestial globe which has a similar set of lines.

The lines running round the star globe parallel to the celestial equator are spaced uniformly in angular distance from the centre of the sphere in degrees, and are called 'lines of declination'. Lines to the north of the celestial equator are written with a + sign or given a suffix N. Those to

The Celestial Sphere

Fig. 2.1

the south of the celestial equator are written with a — sign or have a suffix S. The equator has a declination of 0°.

The lines running north to south are called lines of Right Ascension. These are for convenience spaced at intervals of 15° to mark on the globe the 24 hours of sidereal time, starting from the one numbered 0° which passes through the special point ♈ where the celestial equator crosses the plane of the ecliptic at the instant of the 'spring equinox' when the Sun's declination is zero. This is a convenient point for starting the marking of the 'Right Ascension' of stars. The RA of ♈ is 0° or 0 hours. The plane of the ecliptic is the plane in which the Sun appears to move on the celestial sphere. It is tilted at an angle of 23°.44 with respect to the plane of the celestial equator. This is best understood by referring to a star globe.

The lines of Right Ascension and declination on a star globe thus provide a reference system for all celestial objects, just as terrestial longitude and latitude provide a reference system for places, ships or aircraft on the Earth.

To complete our model we have to imagine the Earth globe as a small speck turning from west to east on its axis once every 24 hours and situated at the centre of the celestial sphere (see Fig. 2.1). This turning creates the illusion that the celestial sphere is turning east to west, and gives us our diurnal time system. The Earth turns at about 15° every hour with *respect to the Sun*, but apparently 15°.04108 with *respect to the stars*. The apparent addition of 0°.04108 per hour is accounted for by the fact that the Earth is orbiting round the Sun at approximately this angular speed, and in the same direction. The zero RA line (RA = 0°) remains fixed in the celestial sphere and so appears to transit our meridian every 23^{hrs} 56^{min} 4^{sec}. As we have seen, the 0° RA cuts the celestial equator at a point called the 'vernal equinox' which can be considered as the hour hand of a gigantic celestial sidereal clock (see Figs. 2.5 and 2.6). If we know the position of this sidereal hand at any time we know the relative positions of all the stars in the celestial sphere as seen from the revolving Earth.

The Sun plays a double role. It provides us with a convenient means of marking our 24 hour spin and regulates our daily lives by our 24 hour clock, but, because we also move in a near circular orbit round the Sun we get a slightly different view of the stars each month and it gives us our annual four seasons and helps us to mark our year.

The Moon comes into the picture with its waxing and waning approximately every 28 days and so gives us the Month and provides us with the basis for a calendar. The monthly calendar and the solar year and the Earth's daily spin, however, were found not to fit into a tidy system of whole numbers and caused endless discussion and controversy among early astronomers who tried to reconcile the apparent anomalies of the calendar.

2.3

Apart from a star globe, another valuable part of a budding astronomer's stock in trade is a good almanac, or an astronomical Year Book, which lists all the easily visible stars in order of their Right Ascension, and gives their declinations. For astro-navigation the *Nautical Almanac* which is produced jointly by H.M. Nautical Almanac Office, Royal Greenwich Observatory, and by the Nautical Almanac Office, United States Naval Observatory, is essential for accurate work.

A table is given each month in most almanacs or astronomical Year Books telling you just how the Earth is situated with respect to the Sun, stars and planets in the celestial sphere at any time of that month or at any hour of the day; it thus tells you what part of the celestial sphere you can see at any time.

For example, on 20th November 1976 we were told that at 0^h the Sun was at RA 15^h 42^m 07^s and declination $-19°40'$ so we can mark this point on our globe. We would not of course have been able to see any stars in that region of the sky at that time because the Sun was occupying this region, but some 12 hours from 15^h 42^m 07^s we would have been able to see stars

The Celestial Sphere

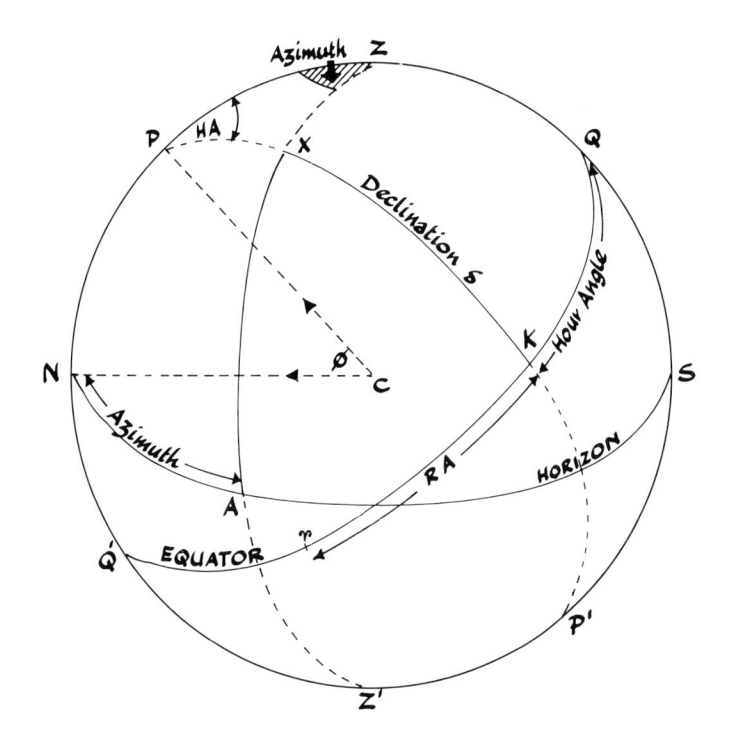

Fig. 2.2

P is the North Celestial Pole.
P' is the South Celestial Pole.
Z is the Zenith.
Z' is the Nadir.
X is the Star or Celestial Body.
ZPX is the Spherical Triangle we consider.

$\angle ZPX$ the Hour Angle (HA) = KQ.
XA = the Altitude and ZX = co-altitude or Zenith Distance.
XK = the Declination, δ, of the star (i.e. angle off the equator).
ΥQ = Local Sidereal Time = Hour Angle* + (RA)*.
QK = Local Hour Angle* = $\Upsilon Q - \Upsilon K$.
ΥK = Right Ascension of Star (RA).
NP = Latitude of observer, ϕ.
PZ = Co-latitude of observer $(90-\phi)$.
$\angle PZX$ = (NA) = Azimuth (Az).
$PX = 90-\delta$ = North Polar Distance or co-declination.
$ZQSP'Z'NP$ is the Observer's Meridian Plane.

When a star is culminating, i.e., it is in the meridian plane, then its hour angle is zero, and the local sidereal time at that instant is equal to the Right Ascension of the star.

13

in the southern sky with RA 3^h 42^m 07^s, i.e., Murfak, The Pleiades, Aldebaran and the stars of Orion.

Again we find on this date that the sidereal time at 0^h was 3^h 56^m 32^s. This tells us where the crossing of the ecliptic and the equator at the vernal equinox happened to be at that time on the sidereal clock.

The equator and the ecliptic (which can be regarded as the path on the celestial sphere followed by the Sun), necessarily cross at two points, one of which we have already considered, known as the vernal equinox, and the other one known as the autumnal equinox and denoted by the symbols ♈ and ♎ respectively. They are separated by 180°. At these points the Sun's declination is 0° and can readily be identified on a star globe. We choose the vernal equinox ♈ as the big hand of a star clock or sidereal clock and the vernal equinox marks the zero from which the Right Ascension of stars is measured on the celestial sphere.

Figure 2.2 represents the celestial sphere, and shows the various terms in current use in positional astronomy and astro-navigation. It can be used to maximum advantage in conjunction with a star globe (see also Fig. 3.14).

2.4 The Planisphere

A representation of the stars and celestial bodies showing their positions at any time is very effectively provided by means of a 'planisphere'. This shows on a flat surface the stars with their polar coordinates of Right Ascension and declination (see Fig. 2.3). Star constellations are necessarily slightly distorted on a planisphere as it is not possible to represent perfectly on a plane surface a part of the celestial sphere, although the coordinates of Right Ascension and declination can be accurately mapped.

The planisphere disc shows the Right Ascensions of all stars in hours as well as in degrees, measured anti-clockwise from the equinox, ♈. The disc also shows the ecliptic or the path of the Sun in its apparent annual journey through the stars and the dates on which it reaches each point in the ecliptic. The dates are inscribed on the rim of the disc.

Over this disc is placed a second concentric disc or screen, Fig. 2.4, which is opaque (except for an elliptical hole, *H*, and is divided round its perimeter into 24 parts, numbered 0 to 24. This is in effect a 24 hour clock face having the 0/24 hour mark labelled north, the 12 mark labelled south, the 6 hour mark labelled east and the 18 hour mark labelled west. The two discs are free to turn about an axis through their centres, i.e., through the North Pole and the clock face centre.

The Celestial Sphere

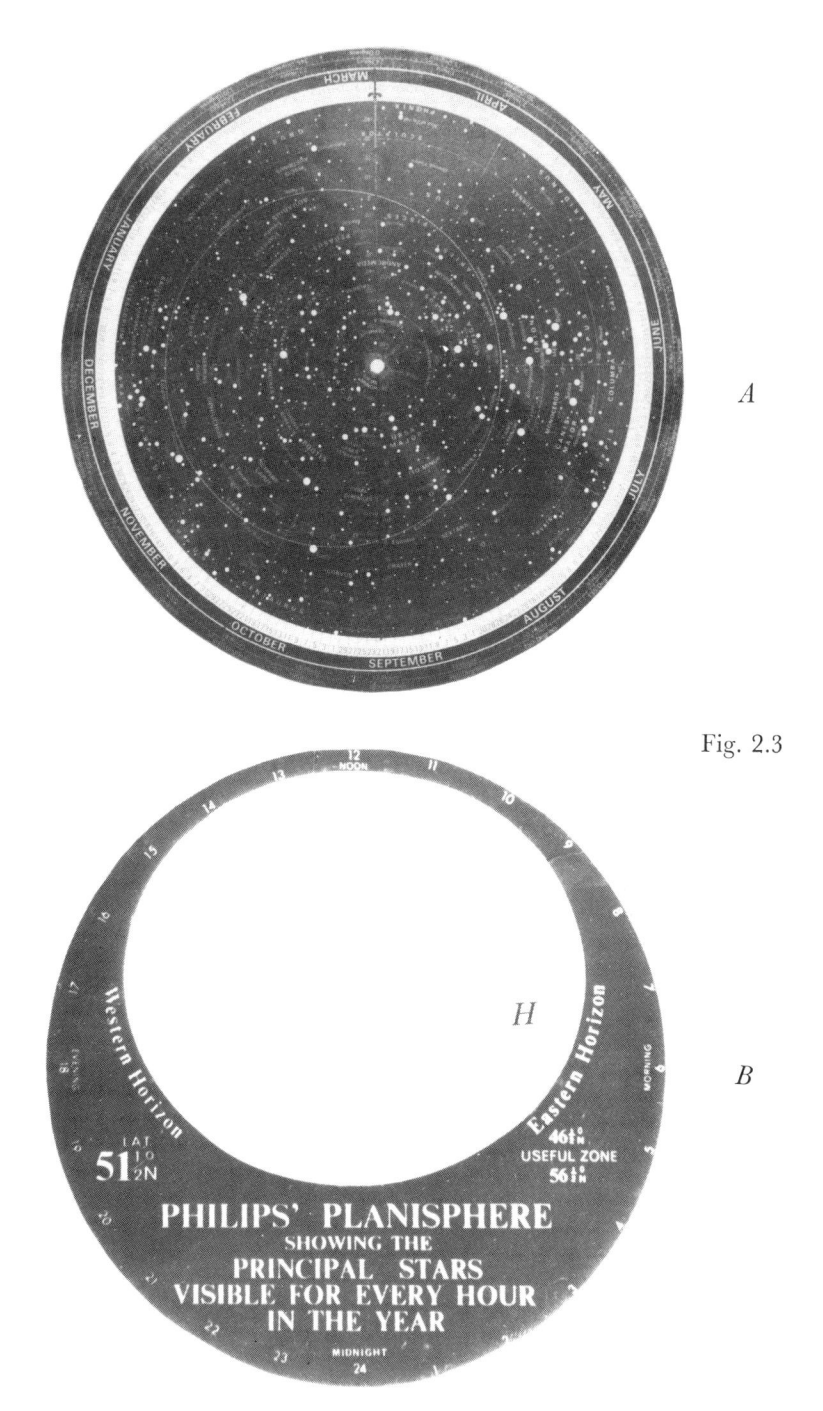

A

Fig. 2.3

H

B

Fig. 2.4

15

The elliptical hole, *H*, cut in the superimposed disc shown in Fig. 2.4, has its centre on the 51° declination mark as this is the latitude for which this particular planisphere is designed. The hole discloses that part of the heavens visible to an observer at any time.

It is a useful exercise to make a planisphere from stiff paper, preferably polar graph paper as this facilitates the plotting of star positions.

The planisphere is an excellent device for showing how stars appear to move ahead of the Sun by about 4 minutes (or 1°) every day. This steady advance is well demonstrated by the turning of the planisphere star disc in an anti-clockwise direction, but keeping the top disc fixed.

The equinox, ♈, opposite 21st March, as we have seen, can be regarded as the hour hand of a 24 hour clock, and the time it reads is the *sidereal time*. This sidereal time also shows the Right Ascension of stars that are culminating *at that time*. It follows that if we set the date on the planisphere disc on the local time of the clock disc we shall have the proper setting of the planisphere showing the local hour angle of any particular star, i.e., the angle between the observer (on the meridian), the pole and the star, or ZPX in Fig. 2.2.

The photograph shows the positions of stars on 1st December at 2000 hrs or the positions of stars on 17th September at 0100 and in fact positions on all dates at corresponding times.

The principles upon which the planisphere is based suggest a simple way of finding approximately in terms of the local time, when a particular star will transit the meridian on a particular date.

Beginners may encounter a difficulty here because the imaginary hand of the sidereal clock ♈ gives sidereal time 0^h when the sidereal hour angle is 0°, that is when the 'hand' is due south. When however the Sun registers 0° hour angle the local sun time is 12^h and the *Sun* is due south.

On 21st March, the Sun and the vernal equinox ♈ occupy roughly the same position on the celestial sphere and show the same hour angle. The almanacs and ephemerides will then show that when the sidereal time is 0^h the local time is 12^h, i.e., when the Sun transits the meridian at 12 noon local sun time, the equinox is showing hour angle 0°. On the 21st September however, the equinox and the Sun are 180° or 12 hours apart so that for this particular day the sidereal time and the local sun time have approximately the same value. The ephemerides will show that on this date sidereal time at 0^h local time, is also 0^h. For each day after 21st September the sidereal clock will gain approximately 4 minutes, or 1° of hour angle. (The more accurate gain per day is $3^m 55^s.91$ of time as given in Section 2.9.)

The Celestial Sphere

This provides a way of finding the sidereal time at midnight on any day of the year, and consequently this gives us the Right Ascension of the star or stars in transit past the meridian at this time.

For example, we may wish to find at what time Sirius, RA $6^h\ 44^m$ will culminate on 16th February. This date is 148 days past the 21st September, and therefore the sidereal time at midnight on this date will be $148 \times \frac{4}{60}$ hours = 9.866 hrs. This means that stars having RA 9.866 hours will culminate then. Sirius with RA 6.733 will culminate $(9.866 - 6.733)$ hours = 3.133 hours before midnight, i.e., at $(24.00 - 3.133)$ hours = 20.867 hrs or $20^h\ 52^m$.

This result can readily be checked by means of the planisphere.

Although this method of finding a star's position at any time is rough and ready, it is sufficient to provide a means of telling you which stars are likely to be visible and when they culminate on any day of the year.

A ready reckoner giving the number of days past 21st September, of any particular date can be easily made and kept for reference by numbering the days consecutively on an old calendar commencing with 22nd September *(see also Appendix IV)*.

2.5 The planisphere can be regarded as a device which connects the three fundamental parameters of elementary positional astronomy:

(1) The local mean time.
(2) The date.
(3) The local hour angle.

It follows that if we know any two of these we can obtain the third from the planisphere.

In normal use we know the local time (1) which is set opposite the known date (2) and we then observe the position or the local hour angle (3) of the star we are interested in, and so identify it.

It is an instructive and useful project to use the planisphere as a nocturnal, that is as an instrument now mainly of historical interest, for telling the time from the positions of the stars. This is a matter of considerable interest for young astronomers, Boy Scouts and youth leaders.

To use the planisphere as a nocturnal, observe the position, i.e., the hour angle (3), of a known star, as described below. Then set this position on the planisphere opposite the date (2) and read the required local time (1).

This procedure will give only a rough estimate of the local time, so for greater accuracy we select two stars that are circumpolar for northern latitudes and having approximately the same hour angle. The two stars in Ursa Major (the Dipper or Plough), Merak and Dubhe, known as the pointers, are ideal for this purpose. They each have a Right Ascension of approximately 11 hours.

As an aid to an accurate positioning of these two stars on the planisphere a thin knitting needle P about 17 cm long is attached by a suitable adhesive to the back of the disc, A (Fig. 2.3). One end should be just clear of the central hole in the discs and the needle should lie along the line joining the central hole (the pole) and the edge of the date circle opposite the date 6th September or the Right Ascension mark 11 hours. This ensures that the needle is in the same line as the pointers of the Plough or Dipper, Dubhe and Merak. A 25 cm transparent plastic ruler R is now stuck on the front of the 24 hour clock disc, B, Fig. 2.4 so that it lies with its centre line along the line joining the 12 noon mark to the 0^h (12 midnight) mark, with about 6cm of ruler projecting to serve as a convenient handle (see Fig. 2.6).

To use the planisphere-nocturnal, hold it about 20 cm from the eye with the Pole Star visible through the centre hole of the planisphere disc, which is held perpendicularly to the line of sight of the Pole Star. Rotate the knitting needle until it is in line with the two pointers, being careful to keep the ruler in the plane of the meridian and perpendicular to the polar axis. The star chart is now correctly aligned to show the positions of all visible stars. Use a hand torch to locate the date on the date circle and note the time opposite the date. This time* is the required local mean time for the observer.

2.6

Figure 2.6 in conjunction with Fig. 2.5, the planisphere, will help to give the precise relations between a star's local hour angle, its RA and the local sidereal time relevant to the observer. If the observer is on the Greenwich meridian the local hour angle and the local sidereal time become the 'Greenwich hour angle' and the 'Greenwich sidereal time' respectively.

The local sidereal time is the interval in sidereal hours, minutes and seconds that has passed since the 1st point of Aries (vernal equinox) (Υ) was last on the observer's meridian. Measured by $AC \, \Upsilon$ (Fig. 2.6).

The Right Ascension of a star or celestial body can be considered as the time that elapses between the passage of Υ, the equinox, across the meridian and that of the body. Measured by $SC \, \Upsilon$ (Figs. 2.5 and 2.6).

*It is important to remind beginners that the local mean time arrived at by the nocturnal must be corrected for the observer's longitude to obtain the Standard Zone Time.

The Celestial Sphere

P

R

Fig. 2.5

Fig. 2.6

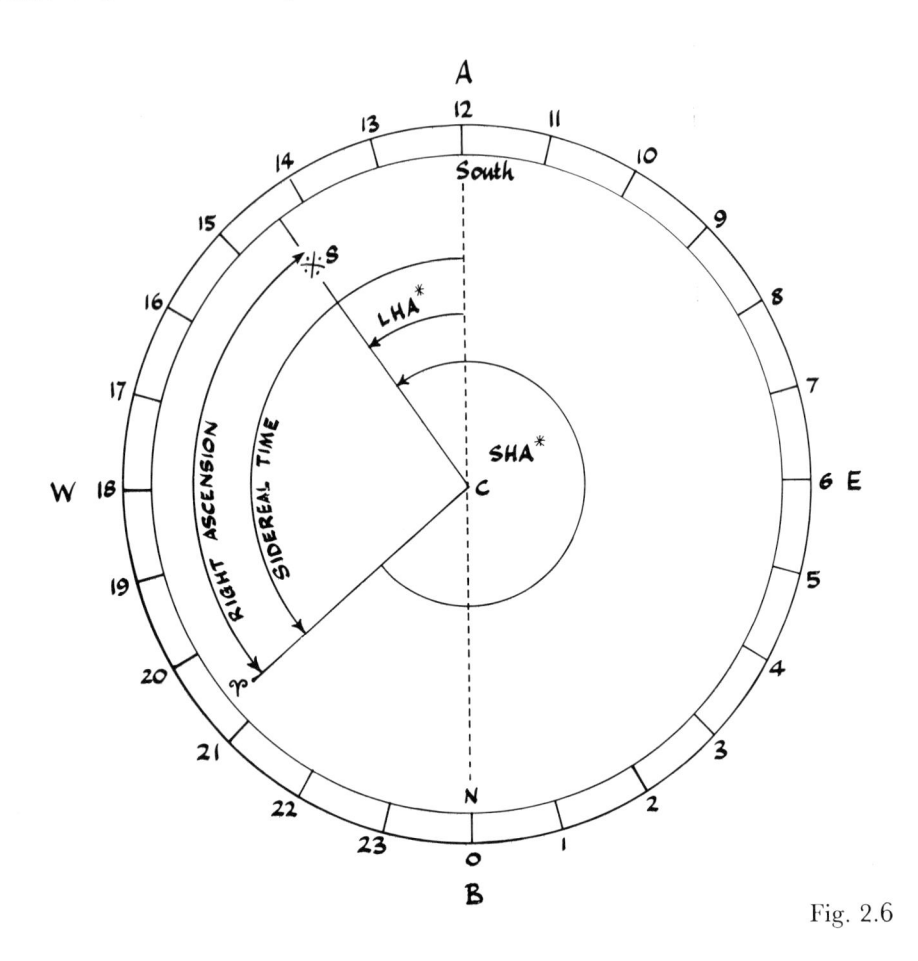

Draw a line through the centre, C, ACB to represent the NS meridian, with GMT noon at the top. Then a star, S, at some particular time makes an angle ACS and this is the star's hour angle at that time.

We now have the following relations from Fig. 2.6:

$$AC\Upsilon = \text{local sidereal time}$$
$$ACS = AC\Upsilon - SC\Upsilon = (\text{LHA})^*$$
$$(\text{LHA})^* = LST - RA^* \qquad [2.6(1)]$$

We are given sidereal time in *Whitaker's Almanack* for 0^h GMT each day or in the *Nautical Almanac* or in the *American Ephemerides* at intervals of 1 hour throughout the day.

The Celestial Sphere

2.7 The Sun's Changing Position on a Planisphere

The position (RA and declination) of the *Sun* is not marked on the star disc of the planisphere as it varies with the date. It can be found from *Whitaker's Almanack* or one of the ephemerides. The Sun in fact lies on the Right Ascension line drawn from the pole to the date mark on the disc, at the Sun's declination for that day. The RA of the Sun thus moves *roughly 1°* per day from 0° on March 21st, to 180° on September 21st and then on to the 360°/0° mark on March 21st of the next year. The Sun thus slips away from ♈ a total of 360° in 365½ days during the course of a full year.

The position of the Sun on the planisphere, Fig. 2.5, *with respect to the 24 hour clock face*, marks the mean time of day, so when the date on the star disc (really the Sun's position) is brought into coincidence with the time on the clock disc, B (i.e., sun time), the planisphere is set for the heavens as they appear on that date and at the time of day indicated by the date, as mentioned in Section 2.5.

The relation

$$(LHA)^* = \text{local sidereal time} - (RA)^*$$

from [2.6(1)] above is applicable to any celestial body, including the Sun, but the RA of the Sun is subject to a fairly rapid daily change arising from a number of factors, chief among which are the Sun's apparent irregular orbital habits, and so an alternative expression is used for the LHA true sun as follows.

The compilers of nautical almanacs have saved astronomers and navigators a great deal of trouble by making a 'package' of most of the Sun's irregularities under the cryptic symbol E known as the 'equation of time' which is something the calculator can, as a matter of interest, calculate from basic data in a few seconds (see Chapter 8).

E is an interval of time used to denote how much the true sun is ahead or behind in time of local mean time. The LHA relation for the true sun is,

$$(LHA) \text{ true sun} = GMT \pm 12 + \text{equation of time} \pm^E_W \text{ Long} \quad [2.7(1)]$$

The 12 appears because the HA of the Sun at 12 noon sun time $= 0$.

2.8 Sidereal Hour Angle

It should be noted that *Whitaker's Almanack* and many astronomical handbooks gives the RA of bodies, but the *Nautical Almanac* works the other way round and gives the sidereal hour angle of a star, which is simply

360° − (RA) or (SHA) = 360 − (RA). It follows from Section 2.4 that,

$$(HA)^* = (GHA)\varUpsilon + (SHA)^*$$

This is a useful relation to use if you have a *Nautical Almanac* because the GHA ♈ (or the *sidereal time*) is listed for every hour of the day, and for every day of the year, and there are further sets of tables which give the necessary increments to be added for the minutes and seconds between the hours. The sidereal hour angles of 57 of the more important visible stars are given on each page opening, so that finding a star's GHA at any time is a simple addition sum.

For example, in the entry at 22.00 GMT on 24th November 1977, the Greenwich sidereal time ♈ is given as 33° 44.4′ under the column 'ARIES GHA' which is another way of saying that the Greenwich sidereal time then is 33° 44.4′ or $2^h.24933$ or $2^h\ 14^m\ 58^s$.

2.9 Reducing Sidereal Time to Mean Time

We have seen that the sidereal hour angle of a star—used in the *Nautical Almanac*—is a measure of how far a star is ahead of ♈ in the rotating celestial sphere. The Right Ascension is a measure of how far a star is behind ♈ in the 360° circle of the celestial equator.

From *Whitaker's Almanack* on page 2 of each month under the column headed 'Sidereal Time', is given the sidereal time at 0^h UT for each day, so in order to get the sidereal time at T hours (UT) we have to add to sidereal time at 0^h a further T hours. But as we have seen in Section 2.2, this will give us only an approximate result because during these T hours of mean time, ♈ gains on the Sun at the rate of approximately 10 seconds every hour. More accurately it can be calculated that a sidereal clock gains on the mean solar clock at the rate of $3^m\ 55^s.91$ every 24 hours of sidereal time or $3^m\ 56^s.56$ every 24 hours of solar time.

This result can be arrived at simply as follows: the Earth during one tropical year (equinox to equinox) turns on its axis 366.242 2 times, but because of its revolution round the Sun, it appears to turn only 365.242 2 times with respect to the Sun. A mean solar clock keeping UT runs slower than a sidereal clock by the factor,

$$\frac{366.242\ 2}{365.242\ 2} = 1.002\ 738$$

Using this approximate factor it is easy to reduce sidereal time to solar mean time and vice versa.

The Celestial Sphere

Let the two clocks start together at $00^h.00$ UT. Then after 24 hours of mean solar time, the mean time solar clock will of course read 24 hours, while the sidereal clock will read $24 \times 1.002\,738$ sidereal time $= 24^h.065\,712$ or $24^h\,3^m\,56^s.56$. The sidereal clock will thus have gained $3^m\,56^s.56$ during the 24 hours of mean sun time.

When however the sidereal clock shows 24 hours of sidereal time, the mean sun time clock will show

$$\frac{24}{1.002\,738} \text{ hours} = 23.934\,467\,43 \text{ hours}$$
$$= 23^h\,56^m\,4^s.09$$

The sidereal clock has gained $3^m\,55^s.91$ over the mean sun clock during 24 hours of *sidereal* time.

We therefore have 24 hours of mean solar time $= 24^h\,03^m\,56^s.56$ of mean sidereal time, and 24 hours of mean sidereal time $= 23^h\,56^m\,04^s.09$ of mean solar time.

For a more comprehensive account of sidereal time and systems of time measurement, reference should be made to the Explanatory Supplement of the *Astronomical Ephemeris* and *Nautical Almanac* prepared by the Nautical Almanac Offices of the UK and the USA.

So using sidereal time at 0^h from *Whitaker's Almanack* we can write:

sidereal time at T hours (UT) = sidereal time at 0^h(UT) $+ T \times 1.002\,738$

$\therefore \quad$ (GHA)* = sidereal time at $0^h + T \times 1.002\,738 - $(RA)*.

This is correct for the longitude of Greenwich, but we can easily shift the situation to say longitude W° or longitude E°, as the effect of this shift is to decrease the hour angle if west or increase it if east. So we have

Local hour angle = sidereal time at $0^h + 1.002\,738\,T - (RA)^{+E}_{-W}$ Long

It is this local hour angle that is commonly used in positional astronomy and in astro-navigation.

We are now able to calculate the hour angle of the equinox ♈ at any time simply by making use of any almanac. For example, the sidereal time at 0^h on 24th November 1976 was $4^h\,12^m\,19^s$. It is convenient to work in hours and decimals of an hour when using a calculator. The sidereal time $4^h\,12^m\,19^s$ is rapidly reduced to $4.205\,277$ hours by the calculator without any paperwork, as follows:

$$4^h\,12^m\,19^s = \frac{(19 \div 60) + 12}{60} + 4 \text{ hours}$$

and displays the result $4.205\,277$ hours.

23

Now to get the sidereal time at 22.00 GMT we add to 4.205 277 the sidereal time equivalent of 22.00 hours or $22^h \times 1.002\,738$ as above, giving sidereal time at 22.00 GMT $= 4.205\,277 + (22 \times 1.002\,738)$. The calculator displays in a few seconds 26.265 5 hours which is equivalent to 2.265 5 hours (by subtracting 24 hours or 360°). Multiply this by 15 and we have Greenwich sidereal hour angle of the equinox, in degrees ♈, at time stated above, $= 33°.98$ or $33°58'.8$ which agrees exactly with the value given in the *Nautical Almanac* for this date and time in the column Aries GHA.

We have found the sidereal time on 24th November 1976 at 22.00 UT both by *Whitaker's Almanack* and also by the *Nautical Almanac* and found it to be 33°.98 or 33°58'.8. The next stage is to get the Greenwich hour angle: e.g. of Betelgeuse

(1) By Whitaker, which gives the mean RA of Betelgeuse, $5^h\,53.9^m$ or 88°28'.5. Hence,

$$(GHA)^* = \text{sidereal time} - (RA)^*$$
$$= 33°58'.8 - 88°28'.5 \quad \text{(add 360° to keep result positive)}$$
$$= 305°30'.3.$$

(2) By the *Nautical Almanac*, which gives SHA Betelgeuse 271°31'.7 for 24th November 1976:

$$(GHA)^* = \text{sidereal time} + (SHA)^*$$
$$= 33°58'.8 + 271°31'.7$$
$$= 305°30'.5.$$

This is 0'.2 greater than the value from *Whitaker's Almanack* and can be explained because *Whitaker's Almanack* gives only a mean value for the RA*.

It is satisfying to find that the GHA worked out by the use of *Whitaker's Almanack* agrees to within a fraction of a 1' with that computed by the *Nautical Almanac*.

Most star watchers have a meridian line marked in their garden. This is indispensable for those who wish to indulge in practical work in positional astronomy. This marking should be made accurately and fixed by two straight vertical rods about 2 m high and at least 5 m apart so that a string joining them casts a shadow exactly in line with the shadow of the rods when the Sun transits the meridian, i.e., at the instant in GMT 12+longW (or −longE)−equation of time. A weight of about 1 kg hung from the middle of the string by a string about a metre long and immersed in a bucket of water to damp down the tendency to swing will

The Celestial Sphere

help in making an accurate marking. To determine the exact GMT at which a planet or star crosses the meridian we use the fact that at transit the local sidereal time is then equal to the planet's Right Ascension. The GMT of a transit can be calculated as follows:

Let GMT of transit be T in hours and decimals of an hour.

At transit   local sidereal time $=$ RA   (star or planet).

But   local sidereal time $=$ sidereal time at $0^h + T \times 1.002\ 738\ {}^{+E}_{-W}$ Long.

$\therefore$ (RA) $=$ sidereal time at $0^h + T \times 1.002\ 738\ {}^{+E}_{-W}$ Long

or T time of transit $= \dfrac{(RA) - ST \text{ at } 0^h {}^{+W}_{-E} \text{ Long}}{1.002\ 738}$

To find the time of transit of Venus on, say, 2nd May 1977 and its altitude at transit. This is a good way of spotting Venus in broad daylight. From an almanac we find on that day that RA Venus is $0^h\ 28^m\ 56^s$. Suppose we are in longitude $1° 44'$ W. ($6^m\ 56^s$).

Sidereal time at 0^h on this date is given as $14^h\ 39^m\ 11^s$.

The GMT of transit $= \dfrac{0^h\ 28^m\ 56^s - 14^h\ 39^m\ 11^s + 6^m\ 56^s}{1.002\ 738}$

(using the calculator) $= \dfrac{9^h\ 56^m\ 41^s}{1.002\ 738} = \dfrac{9.944\ 722}{1.002\ 738} = 9.917\ 56$

$= 9^h\ 55^m$

This corresponds to the time of meridian passage for long $6^m\ 56^s$ given in the *Nautical Almanac*.

Next, to find the altitude of Venus at transit:

Obtain the declination δ from *Whitaker's Almanack* or the *Nautical Almanac* and add this algebraically to the complement of the observer's latitude, $90 - \phi$.

Altitude at transit $= (90 - \phi + \delta)$.

In the example the declination of Venus is $+5°24'$. If the latitude of the observer is 51°N, so the altitude at transit is:

$$90 - 51 + 5°24' = 44°24'.$$

We thus have the exact position of Venus on *2nd May at $09^h\ 55^m$ as Az180' (due south) and altitude 44°24'*.

Thus, by facing exactly due south along our garden meridian at $9^h\ 55^m$

and looking at a spot at an altitude 44°24′ (using for example the simple cross staff or astrolabe described in Section 3.2 of the next chapter) the planet should appear to the naked eye or through a pair of binoculars held steadily in position.

Before proceeding to spherical triangles readers could, with profit, study a few more examples in that part of astronomy that involves only judicious adding and subtracting of those elements of a celestial body's position found on the star globe which are dependent on the time and the Earth's rotation. Positions fixed by altitude and azimuth, which are often the most convenient and natural way of doing it, cannot be worked out by arithmetic alone.

2.10 Importance of the Hour Angle

The most important combination of *time* and *Right Ascension* from the point of view of a possessor of an equatorial telescope, is the 'local hour angle,' which used with a star's declination enables an observer to bring the star promptly into view and keep it in view if the telescope is driven by a sidereal clock. The hour angle of a star increases steadily with sidereal time, maintained by the observatory sidereal clock, and the star's HA is 0° when the star is on the meridian. At that instant the sidereal time is equal to the Right Ascension of the star.

The hour angle of the Sun is also of fundamental interest to Sundialists, a growing and enthusiastic body of 'home and garden' astronomers of which more information is given in Chapter 7.

Although by far the greatest value of the calculator for astronomers is in dealing adroitly and swiftly with all forms of spherical triangles it is also a great help and time saver, as we have seen, in converting arc (degrees, minutes, seconds) into time and vice versa without the bother of looking things up in tables. It also converts minutes and seconds of time or of arc into decimals of an hour or a degree respectively. Some slightly more expensive calculators perform this operation instantly by a special conversion key.

2.11 Formulae as Calculator Programmes

We shall see in Chapter 3 that a formidable looking expression in spherical trigonometry is no longer an advanced mathematical formula to be

The Celestial Sphere

worked out, using several sheets of paper and tables, but it is simply a programme to be tapped out on a little pocket keyboard consisting of far fewer keys than on a typewriter. Evaluating an expression containing six trigonometrical functions (see Chapter 3) can now be carried out with the same ease and speed as is shown by a young lady who taps out a receipt for a bill of a dozen items on a cash register.

The professional navigator traditionally works out the answers by referring to two or three different volumes of tables, which have to be entered to the nearest degree of observed quantities, and leaves the navigator the task of making several interpolations to obtain any useful degree of accuracy.

2.12 Importance of Understanding Principles

It would be wrong to advocate a blind tapping out of results on a calculator without an understanding of the principles involved; the essential principles of the spherical triangle are no more difficult than those of plane trigonometry. A calculator is a simple computer and will produce an output-answer strictly in accordance with the input, which must be correctly put in or 'programmed'. It will do only what it is set to do. It does only the donkeywork, not the thinking. An eminent surgeon once said: "The human brain is by far the most effective, efficient and sophisticated computer yet devised and it has the great merit of being mass produced by unskilled labour".

3
Spherical Triangles and Early Devices for their Solution

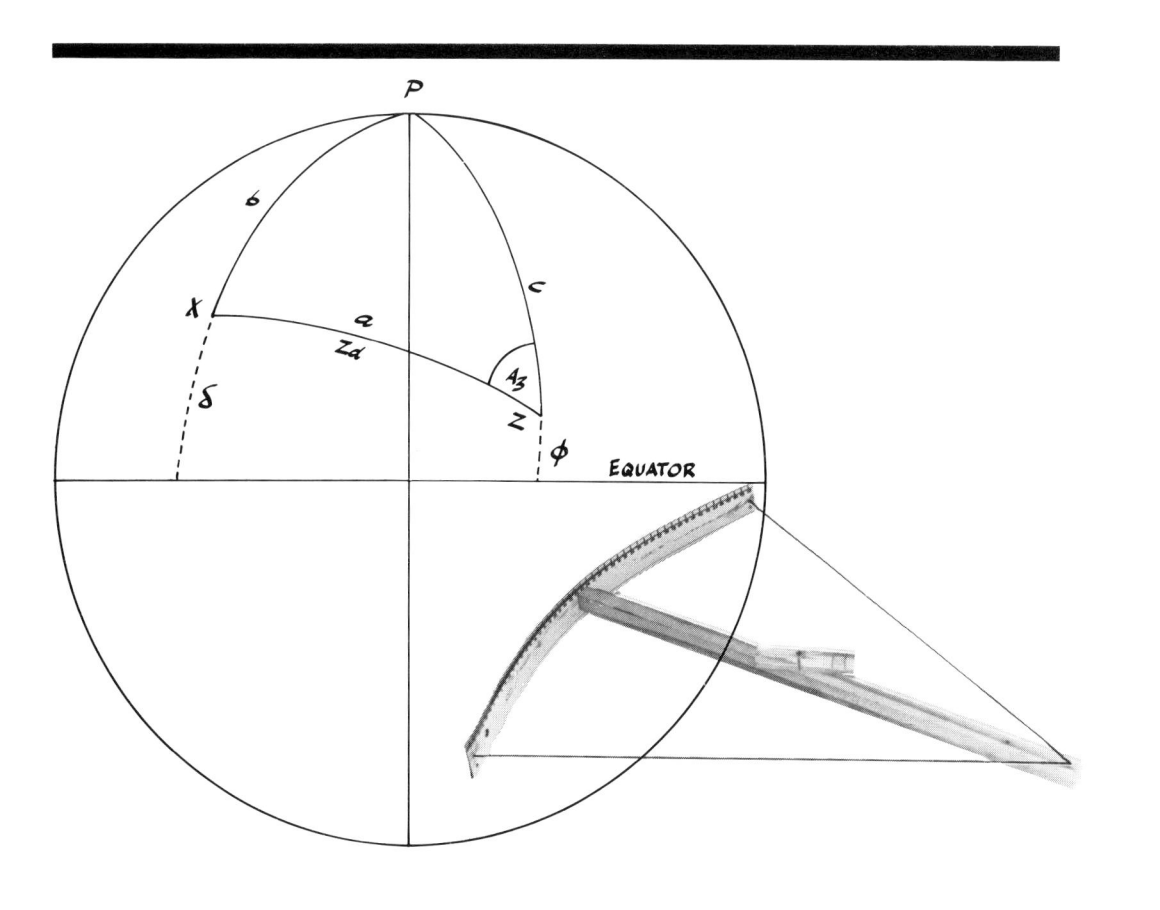

Positional Astronomy and Astro-Navigation Made Easy

> "Most problems in positional astronomy can be solved by the use of spherical trigonometry. The hypothetical concept of a celestial sphere is invaluable as the foundation upon which all fundamental considerations of the positions and motions of heavenly bodies are based".
>
> From the revised version of *Norton's Star Atlas* (Sixteenth Edition), section headed 'Fundamental Concepts'.

3.1 The story of spherical triangles and trigonometrical calculations goes back 2500 years to *Pythagoras*, because his famous theorem about the square on the hypotenuse is the most important single theorem of all mathematics —particularly for astronomers, because it expresses a fundamental characteristic of space in which we live, based on the right-angled triangle which relates the *vertical* and the *horizontal*. Pythagoras thus gave numbers to geometry and thence to astronomy, and this led to trigonometry, which in turn under the Arabs, much later, gave rise to spherical trigonometry. As star gazers they became fully aware that they lived on a sphere within a celestial sphere, and that star positions and patterns could be measured only in terms of angles marked on a sphere.

This development of trigonometry was accelerated by the innovation and spread of Arabic numerals, and an Arabic fascination for calculations. (How they would have loved our calculators!) Astronomers thus became liberated from the restraints of the earlier unmanageable number systems based on Greek or Babylonian symbols.

One problem confronting the early Alexandrian astronomers was how best to represent on a map the stars in their courses across the celestial sphere, at a time when spherical trigonometry was practically unknown— although geometry and trigonometry were well understood and used. Around 300 B.C. *Hipparchus* hit upon the ingenious idea of 'stereoscopic projection', the kind of projection used effectively much later by the Arabs in the making of astrolabes (see Chapter 6).

Much of the mystery associated with the mediaeval astrolabe can be dispelled by the realisation that the astrolabe is essentially a device for 'taking' the altitude of a star. The word 'astrolabe' means star-taker. The *altitude* of a star, sun, moon or planet was all that could be taken. Shakespeare in his Sonnet Number 116 on 'True Love' compares true love with the seemingly fixed Pole Star and ends with the lines:

Spherical Triangles and Early Devices for their Solution

> Love is not love
> Which alters when it alteration finds
> Or bends with the remover to remove:
> Oh no! it is an ever-fixed mark
> that looks on tempests and is never shaken;
> It is a star to every wandering bark,
> whose worth's unknown,
> Although its *height be taken*.

The height above the horizon is there for the taking and a plumbline provides a convenient and perfect line for a reference and forms one all-important side of our spherical triangle, the *co-altitude*. We shall see later how all other sides and angles can follow from this observation, when used in conjunction with the astrolabe. We make early brief mention of the astrolabe because its principles are historically and mathematically basic to our study of positional astronomy, and provide scope for rewarding practice with a calculator.

3.2 Here are six simple devices for taking altitudes of celestial bodies that can be made from ready to hand materials.

A Shadow Astrolabe. The simplest of all for measuring the altitude of the Sun is shown in Fig. 3.1. A project for a lower-school form, and Exercise Number 1 for the calculator!

In Fig. 3.1 *AB* is a strip of wood from a DIY shop, 573 mm × 20 mm × 6 mm. *CB* is a piece of the same kind of wood but with dimensions 106 mm × 20 mm × 6 mm, which is fastened at right angles to *AB* at *B* using a brass corner brace and screws. *BC* = 100 mm. The strip is positioned with *AB* horizontal by means of the spirit level *L*, and points to the Sun, so that the shadow of the top edge *C* of the upright casts a shadow on *AB* at *D*.

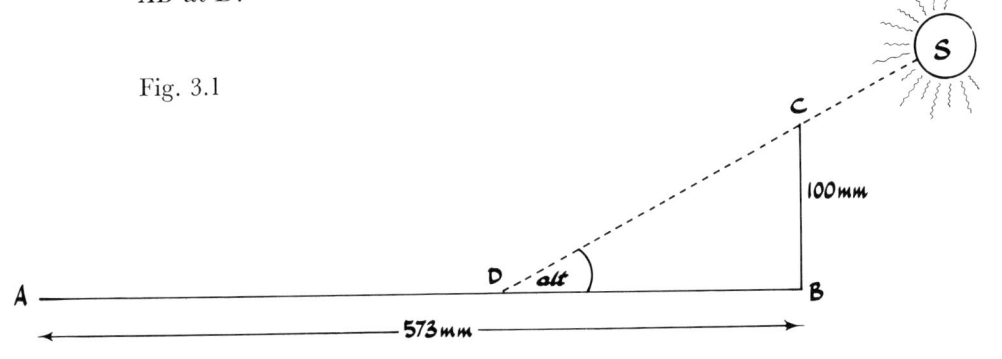

Fig. 3.1

The altitude of the Sun is the angle $CDB =$ (alt) and DB is the length of the shadow cast, d. Then,

$$\tan(\text{alt}) = \frac{100}{d} \quad \text{or} \quad d = \frac{100}{\tan(\text{alt})}.$$

The calculator will give in a matter of minutes a scale for marking AB in terms of Sun's altitudes. Here are a few values for DB or d for various altitudes of the Sun.

Alt. of Sun	d
65°	46.63
55°	70.02
45°	100.00
35°	142.81
25°	214.45
15°	373.20
10°	567.128

The distance d can be marked off from B with a fine pencil or felt pen, for every 5°. (This table was compiled in just under $1\frac{1}{2}$ minutes.)

See Fig. 3.2.

The device can now be used to find the height of the Sun.

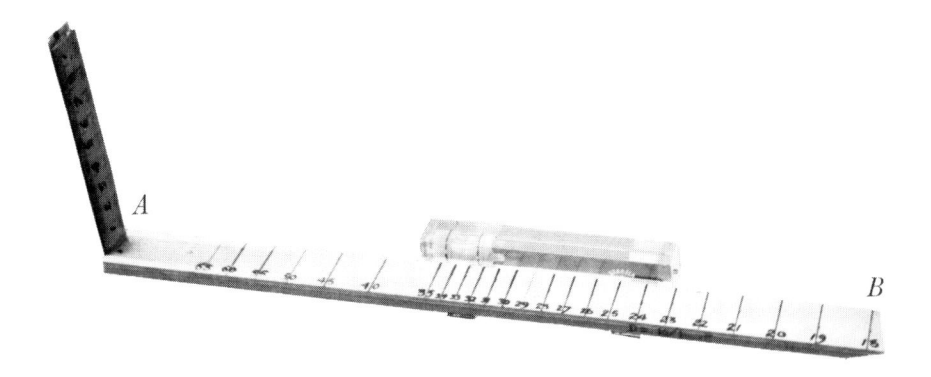

Fig. 3.2. Simple astrolabe for finding the height of the Sun (see Fig. 3.1).

There is a special reason for making the strip $AB = 57 \cdot 3$ cm. Mark the edge of the upright BC into centimetres. Then with the eye placed at A each centimetre on BC will subtend an angle of 1°, because $\tan 1° \approx \frac{1}{57.3}$ (check this on the calculator). Each millimetre on BC will represent $0°.1$ or $6'$.

Spherical Triangles and Early Devices for their Solution

This device is useful for measuring small angles such as the angles between stars, or the heights of lighthouses at sea, or angles between two known landmarks at the observer's position.

The device cannot, however, tell the time directly because, as we shall see, time or the hour angle requires more than an altitude measurement. This simple device is not a sundial but we can use it to tell the local time if we know the declination of the Sun (see Section 7.11), and the latitude.

A Protractor Astrolabe with Glycerine Damping. A simple protractor astrolabe for finding the altitudes of the Sun or stars, and costing only a few pence, is shown in Fig. 3.3.

A is a protractor, 150 mm diameter, graduated in half degrees. *B* is a tin (actually a mustard tin) 60×45 mm containing glycerine to a depth of about 60 mm.

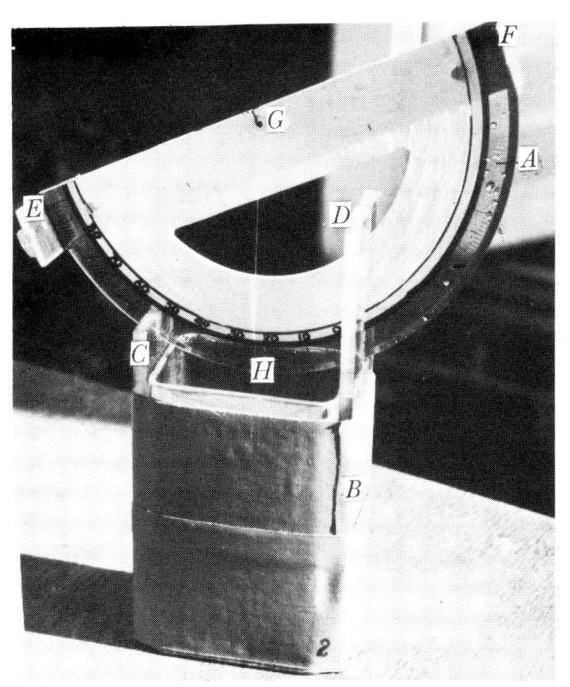

Fig. 3.3. *A*, protractor. *B*, mustard tin. *C* and *D*, perspex supports with slits holding protractor. *G*, centre of protractor. *EF*, sighting bar. *E* is a small piece of a spectacle lens of total length about 5 cm. *F* is a small nick serving as a foresight. *GH* is a plumbline having a lead bob at the lower end which is immersed in glycerine.

Positional Astronomy and Astro-Navigation Made Easy

The protractor is supported in two slits cut in the pieces of perspex C and D as shown in the photograph, so that it can rotate about G, supported on its circumference by C and D. The protractor is fitted with a sighting bar EF cut from a strip of aluminium, and is fitted with a small piece of spectacle lens 165 mm focal length at E to enable the eye to focus clearly on the sight nick F and observe the star at the same time. The altitude of a star is measured by a plumbline, G, of which the bob is immersed in about 200 cm³ of glycerine, which keeps the bob steady and damps out any tendency to oscillate. The motion of the bob is 'dead beat' which makes accurate readings possible.

As soon as a star is in the sights, the instrument is given a slight tilt about the horizontal axis pointing in the direction of the body observed. This tilt causes the plumbline (a thin length of nylon fishing line) to be held to the edge of the protractor at the altitude mark. The glycerine is sufficiently viscous to prevent any slipping of the bob provided the tin is held horizontally and the tilt is about the axis as mentioned above. The plumbline always hangs down the centre of the tin, no matter what altitude is being measured. When finding the altitude of the *Sun*, the protractor is positioned to allow an image of the Sun, formed by a pinhole shown at F, to fall centrally on an etched cross at E. The device can be used even in a wind, and this is a great advantage. An accuracy of 15' can be obtained with this device. The glycerine can be kept in a corked medicine bottle when not in use in the tin.

A 'Cross Staff' Astrolabe. A simple cross staff and back staff type of instrument can be made from a flexible centimetric scale and a spirit level, as shown in Figs. 3.4 and 3.5.

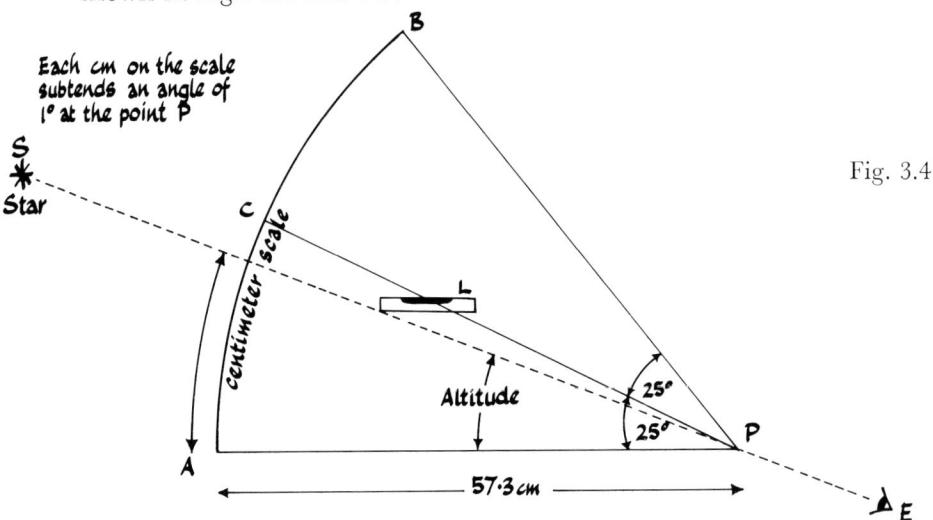

Fig. 3.4

Spherical Triangles and Early Devices for their Solution

In Fig. 3.4 AB is a 50 cm scale with its zero mark at A and the 50 cm mark at B. CP is a piece of wood 2 cm square cross-section which is fixed to the centre of the scale at C by a screw or a small angle brace. At P is a pointed protruberance such as a nail or screw to serve as a backsight. It is 57.3 cm from C, the centre of the scale.

A piece of string or fine wire is secured at one end at A and is taken round the sighting screw P, while the other end of the wire is secured to the end, B, of the scale. The wire is tightened until the conditions shown in Fig. 3.4 are obtained and $AP = PB = 57.3$ cm. Total length of wire $APB = 114.6$ cm. The scale is thus bent approximately in the form of a circle of which the radius is 57.3 cm.

With the eye placed at E a few centimetres from P the backsight, the observer at E has a means of measuring angles between two distant objects or celestial bodies. The range of readings being from $0°$ to $50°$.

Fig. 3.5

To find the altitude of a celestial body it is necessary to ensure that AP is horizontal. This can be done by attaching a small sensitive spirit level L to the wooden piece CP so that the 'level' edge of the spirit level is parallel to the line of sight, PA (where A is the zero mark). This can be checked by observing the horizon from a metre or two above sea level. The backsight, P, the zero mark A and the horizon should be in line as seen by the observer's eye placed a few centimetres beyond P, at E.

When correctly held for observing, light from a star will enter the observer's eye along the path $SCPE$, where C is a point on the edge of the scale. If for example this is observed to be at say the 35.5 cm mark, then the star's altitude is $35°.5$ because each centimetre on the scale subtends an angle of $1°$ at P.

A small mirror 1.5 cm × 3 cm can be fixed or held near the spirit level L to enable the observer to see the bubble in the spirit level and to tilt the bar PC so that PA is horizontal when observing an altitude. A small hand torch is used to illuminate the bubble and the scale when observing at night.

The device can also be used to measure small angles such as heights of cliffs and lighthouses, and when turned in a horizontal position it can measure horizontal angles for checking bearings and compasses.

Sun Sights—the 'cross staff' used as a 'back staff'. The device as described clearly cannot be used for taking the altitude of the Sun, as an observer should never attempt to look directly into the Sun, but it can be adapted for Sun sights by simply turning the device upside down and using it as a 'back staff'.

In this position the device is held so that the Sun casts a shadow of the backsight on to the scale. Then, provided the spirit level is horizontal, the

---

Fig. 3.6 *An Alt–Azimuth Star Finder*. AB is a plastic strip. $PA = PB = PC = 57.3$ cm. $SCPE$ = path of light from star, S. C is the observed point on the scale AB where the star is seen from P. AC in cm = altitude of the star in degrees. L is an accurate spirit level parallel to PA, and Angle $APB = 90°$. G is a wing nut and bolt for clamping the hardboard to the polythene pipe D. The azimuth of a star or the Sun can be measured by the pointer, H, which is fixed to the tube, D, parallel to the plane of the board. The azimuth scale, F, is fixed to the table, G, and is correctly orientated with its 0–180° line lying due north and south.

For observations of the Sun, the board is rotated through 180° and the altitude is measured by the position of the shadow of P on AB, where altitude = APC.

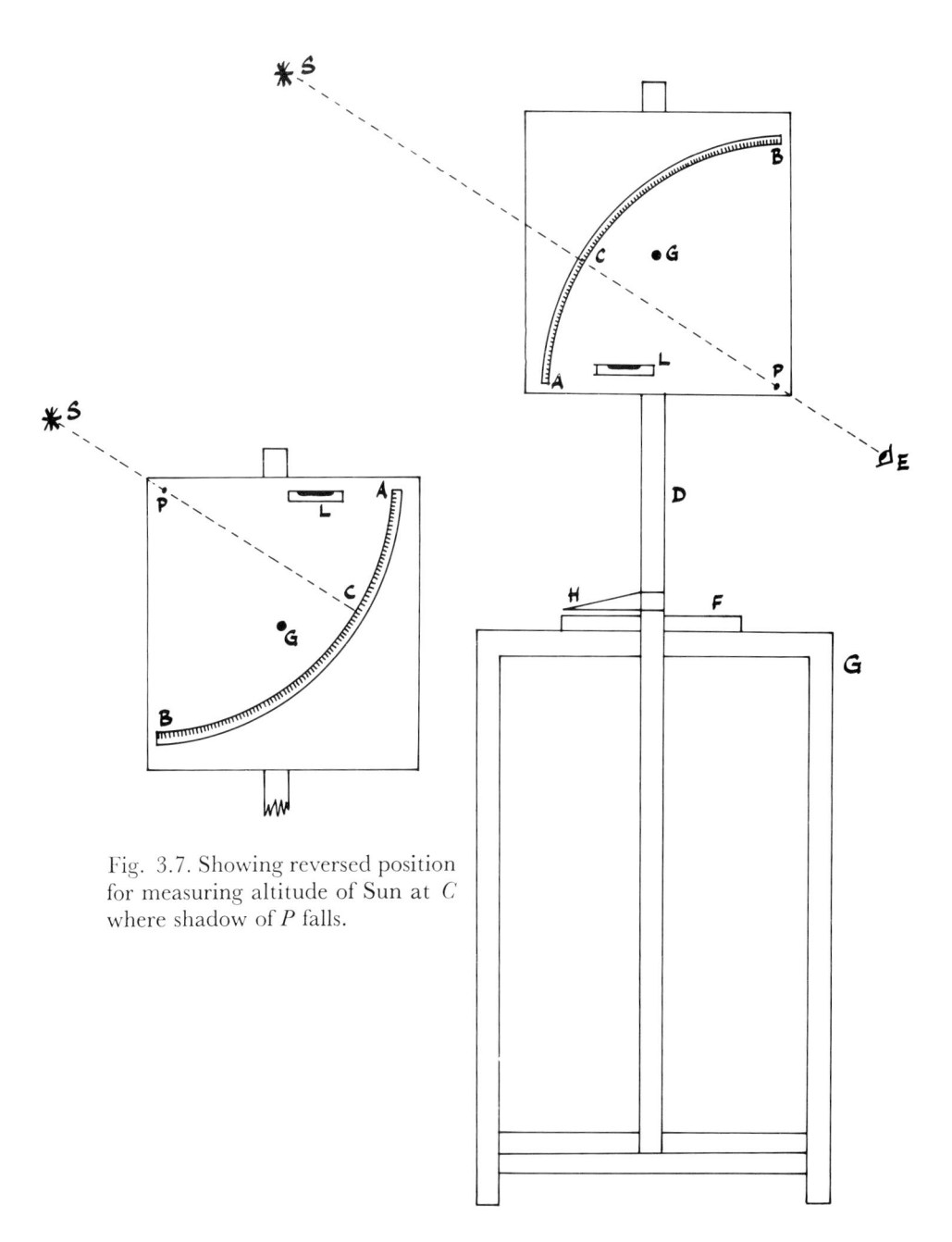

Spherical Triangles and Early Devices for their Solution

Fig. 3.7. Showing reversed position for measuring altitude of Sun at C where shadow of P falls.

Fig. 3.6

37

shadow on the scale will read the Sun's altitude to within 15', when the device is held steadily against a firm support.

The device described in Figs. 3.4 and 3.5 is very simple to make and to use, but nevertheless it provides a quick and easy way of finding altitudes of celestial bodies, and for measuring angles on the celestial sphere, or in coastal navigation.

An Alt–Azimuth Quadrant Star Finder. A more accurate and convenient development of the same principle as that described in Fig. 3.4 is shown in Fig. 3.6. It is a device of very little cost and easy to make. Instead of a plastic ruler a piece of white flexible polythene curtain rail, AB, 92 cm long and about 2.5 cm wide, bent in the form of a quadrant of a circle of 57.3 cm radius, is mounted as shown on a sheet of hardboard about 60 cm square. The curtain rail, AB, is graduated in centimetres and millimetres with 0 at A and 90 at B (a convenient way of doing this is to photocopy a metre scale and use the photocopy markings to stick on the quadrant). P is a small projection at the centre of the quadrant and is the backsight for observations.

In use, AP is kept horizontal using a sensitive spirit level, L. The observer's eye is at E. Suppose the star to be observed is at S, then $EPCS$ are in a straight line, where C is a point on the quadrant, at such a point that APC is the star's altitude, and this is measured in degrees by the distance AC in centimetres.

The firm table supports a vertical tube 50 mm diameter into which is slotted a stout wooden batten to which the hardboard square is attached by a nut and bolt through its centre. This allows a fine adjustment in level as indicated by the spirit level to be made at the time of the observation. It also allows the hardboard square to be turned through 180° into the position shown in Fig. 3.7 which is the position for taking the altitude of the Sun without having to look in the direction of the Sun or to use strong filters. In the reversed position, the position of the shadow of C indicates in degrees the altitude of the Sun, as in Fig. 3.7.

The device as shown can also record or check with an accuracy of a degree or two the azimuth of the body observed on the circular scale on the supporting table. This provides the device with a wide range of uses and projects involving altitudes and azimuths and related spherical triangles.

The device has been found a great help in spotting Venus in daylight, as the altitude can be calculated from the hour angle by relation [3.7(1)] and the corresponding azimuth found.

Spherical Triangles and Early Devices for their Solution

Fig. 3.8(a)

At night a small torch attached to illuminate the scale can be switched on to make the reading as soon as the star and the points *SCP* and *E* are lined up. An accuracy of 0°.1 (1 mm) should be possible with care and skill in observing.

Readings with this device can be taken quickly (this is an advantage in taking the altitude of a satellite) and can be used to measure all altitudes from 0 to 90°. For altitudes greater than about 60° the height of the quadrant can be raised to well above a person's height, or the observer can sit on a chair. Many alt–azimuth optical instruments are difficult to use when observing objects of high altitudes.

Fig. 3.8(b)

Altitudes and azimuths are coordinates to be used and enjoyed. It is worthy of note that the large Northern Hemisphere Observatory Telescope now being planned for the observatory at La Palma, Canary Islands, is an alt–azimuth instrument which will have a mounting of sound engineering design and construction. With modern calculators and computerised controls, the switching from equatorial coordinates to alt–azimuth presents no problem whatever, and a star can be held steadily in the field of view with the same ease and precision as is possible with an equatorially mounted instrument.

Spherical Triangles and Early Devices for their Solution

Simple Home-made Model Sextant. The simple home-made devices for measuring altitudes or angles on the celestial sphere described above cannot of course compare in accuracy with the marine sextant which was invented in the middle of the 18th Century, and then developed until in the early 19th Century, it was established as an instrument of great optical and mechanical precision. A contemporary development was an accurate chronometer which could time an observation to within a second or so, thus bringing about a new era in accurate astro-navigation. For a professional treatment of the sextant, manuals on navigation should be consulted (see Bibliography). Cheap plastic sextants with instruction booklets are now available, and these are accurate to within 0.5 minutes of arc (see Bibliography).

In introductory talks on the sextant it has been found useful to make and use a simple working model of a sextant as a project. A photograph of a model made from ready to hand materials and costing a very little is shown in Fig. 3.9 with the various parts lettered.

- A Index mirror of sextant
- BC Index bar of wood or perspex, able to turn about a small bolt G as an axis. The bar is marked with a fine index line.
- D Horizon mirror about half the area of A and fixed in position.

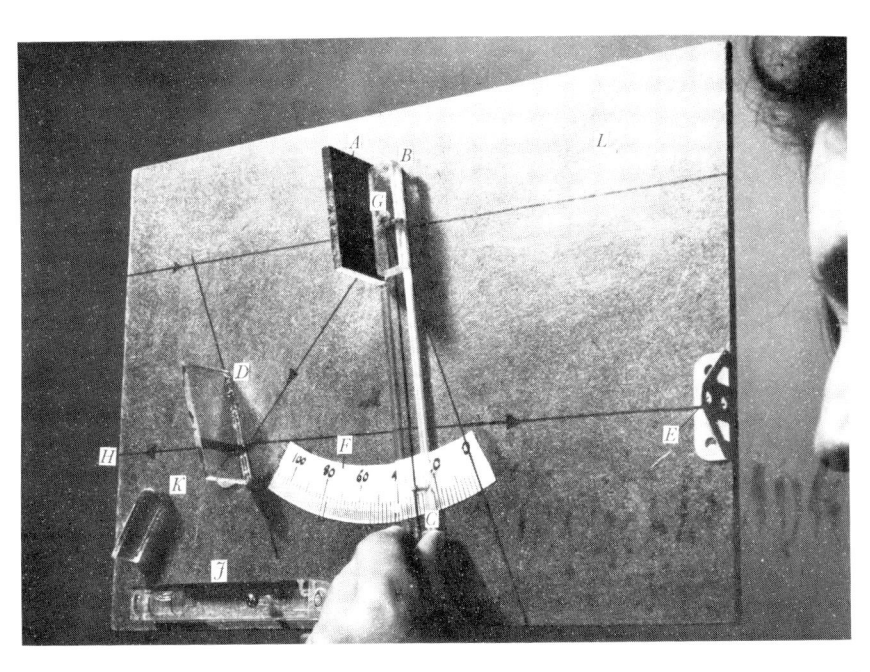

Fig. 3.9

E   Small hole 5 mm diameter for sighting along the horizon line *EH*.

F   A graduated arc of 60°. 1° of arc represents 2° of altitude or angle between two objects.

J   Spirit level as a refinement to ensure that the line of sight is horizontal when the horizon is not visible.

K   Small inclined mirror 30 mm × 15 mm to facilitate view of the bubble of the spirit level from *E*.

L   Base plate of the model, made of hardboard 300 mm × 210 mm, with a handle *N* attached to the back.

The parts listed are mounted on the base board with a good adhesive or small nuts and bolts as appropriate after marking the layout in pencil.

The model is very easy to understand and to use. If made carefully with the mirrors properly aligned just before the adhesive becomes hardened, it can measure angles vertically or in any direction to within 15 minutes of arc.

The scale *F* can be made by photocopying a section of a good large protractor. It will be seen that when the line on the index bar is on 0° on the scale *F*, then the mirror is parallel to the fixed mirror *D*, and when the index arm and index mirror turns through an angle $x°$ then the line of sight from *E* to the reflected object is turned through $2x°$ by the simple laws of the reflection of light.

The model can be adapted to take sights of the Sun, but on no account should the Sun be viewed directly. The dense filters necessary for such a view are provided in marine sextants, but are expensive and impracticable for a simple home-made model. Nevertheless it can be used with reasonable accuracy to measure the altitude of the Sun by holding or clamping the base plate in position in the plane of the Sun's azimuth, and with the horizon line of the sextant horizontal by the spirit level *J*. The index arm is then moved so that the Sun's light is reflected by two reflections, one at the mirror *A*, and the other at the mirror *D*, so that it falls on *E* the viewing hole. To ensure accuracy, a fine black line in 'permanent' felt pen ink is drawn horizontally across the centre of the mirror *A*. This line should coincide with the axis of turning of the sextant mirror.

The index arm *BC* is moved until a shadow of the line is cast across the centre of the viewing hole. No great accuracy is claimed for this model, but it will be as much as the project worker cares to make it by good workmanship and careful alignments.

Marine sextants are very expensive and beyond the means of most schools, but this simple model sextant project may serve as an introduction to the real thing.

3.3 Fundamental Formulae for Spherical Triangles

We saw at the beginning of this Chapter (Section 3.1), that Pythagoras put numbers into triangles and produced trigonometry. It is worthwhile at this point to see how spherical trigonometry grew out of trigonometry and the need to understand how triangles behave when drawn on a sphere. Although the sides of a spherical triangle are sides drawn on a sphere, they are very special sides. They are all parts of great circles of the sphere, and all great circles have their centres at the centre of the sphere.

Each side of a spherical triangle is the shortest distance on the spherical surface between the two points on the triangle it connects, just as the straight line in a plane triangle is the shortest distance between the two points of the triangle which it joins.

Figures 3.10 and 3.11 will serve to make some comparisons between the simple relations in plane trigonometry and spherical trigonometry. The derivation of

$$a^2 = b^2 + c^2 - 2bc \cos A,$$

and

$$\frac{a}{\sin A} = \frac{b}{\sin B} = \frac{c}{\sin C}$$

by elementary trigonometry is a matter of defining sines and cosines and then drawing appropriate right angle triangles as constructions and applying Pythagoras's theorem several times over.

The same process, only more complicated, produces the analogous relations shown for the corresponding relations for a spherical triangle.*

The derivation of these relations is given in *Science for the Citizen* by L. Hogben and is facilitated by the use of home-made cardboard models as shown in Fig. 3.12.

3.4

In the spherical triangle *ABC*, Fig. 3.11, *A*, *B* and *C* are the angles formed by the sides b and c, c and a, and a and b respectively, as in plane trigonometry, but the side a, since it is part of a great circle, can be measured by the angle it subtends at the centre. Similarly for b and c.

This explains the occurrence of functions such as $\cos a$ where a is a side of a spherical triangle.

Placed as we apparently are at the centre of the vast celestial sphere, which for most purposes is considered to be of infinite radius, we are unable to make any celestial measurements except in terms of angles.

*The point here that may be encouraging for beginners is that there is nothing mysterious about the formulae relating to spherical triangles, as these formulae all stem from the well known theorem of Pythagoras, and once they have been established to one's satisfaction, they can be used with confidence as programmes in positional astronomy or navigation.

Positional Astronomy and Astro-Navigation Made Easy

From Fig. 3.11 it can be shown by means of the models in Fig. 3.12 that
$$\cos a = \cos b \cos c + \sin a \sin b \cos A. \qquad [3.4]$$

This is perhaps the most important relation in spherical trigonometry, and indeed as we shall see, in positional astronomy.

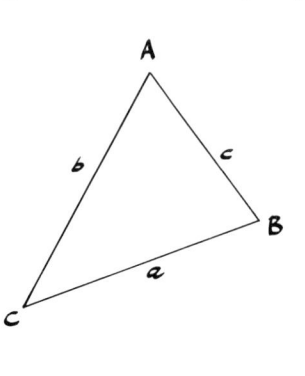

Fig. 3.10

Plane Trigonometry
derived from Pythagoras

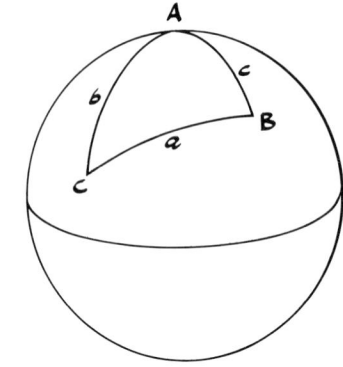

Fig. 3.11

Spherical Trigonometry

Common property:

The sides a, b, and c in each figure are the shortest distances between BC, AC, and AB respectively.

$a^2 = b^2 + c^2$ for right angle at A	$\cos a = \cos b \cos c$ right angle at A
$a^2 = b^2 + c^2 - 2bc \cos A$	$\cos a = \cos b \cos c + \sin b \sin c \cos A$
The sine formula	The sine formula
$\dfrac{a}{\sin A} = \dfrac{b}{\sin B} = \dfrac{c}{\sin C}$	$\dfrac{\sin a}{\sin A} = \dfrac{\sin b}{\sin B} = \dfrac{\sin c}{\sin C}$

3.5 If we now draw a triangle (often referred to as the PZX triangle) on the celestial sphere, Fig. 3.13, with P as the celestial pole, X as the star or Sun with declination δ, and Z for the observer's position overhead, then the angle at P replaces A in the cosine formula above in which a is the distance in angular measure between X and the overhead or zenith point Z, a is known as the 'zenith distance', but a more convenient measure for the navigator is the altitude, which is the complement of the zenith distance.

So for $\cos a$ we can write,
$$\cos a = \sin(90 - a) = \sin(\text{alt})$$

b it will be seen is $(90 - \delta)$ and therefore
$$\cos b = \sin \delta, \quad \text{also} \quad \cos c = \cos(90 - \phi) = \sin \phi.$$

Spherical Triangles and Early Devices for their Solution

Fig. 3.12

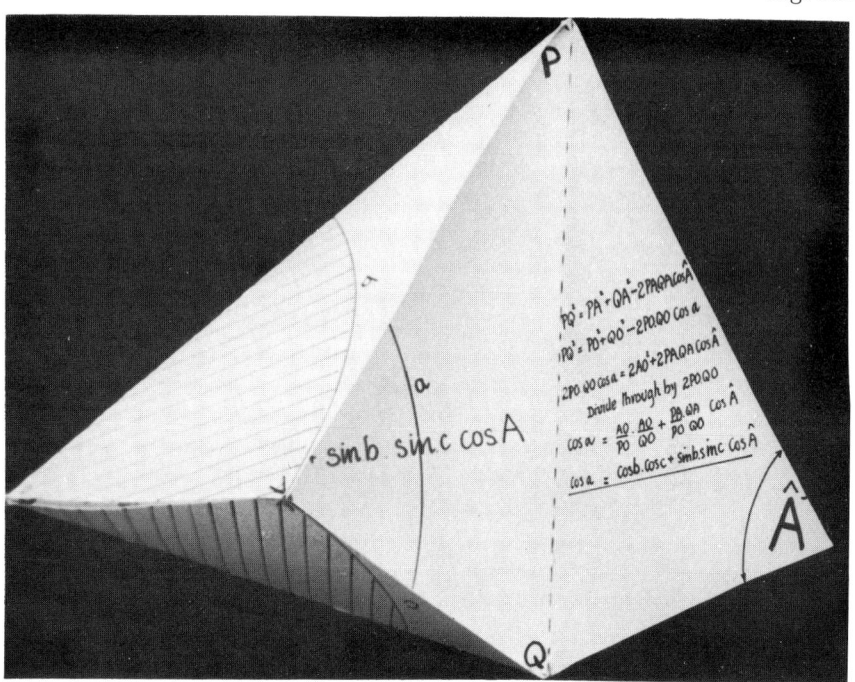

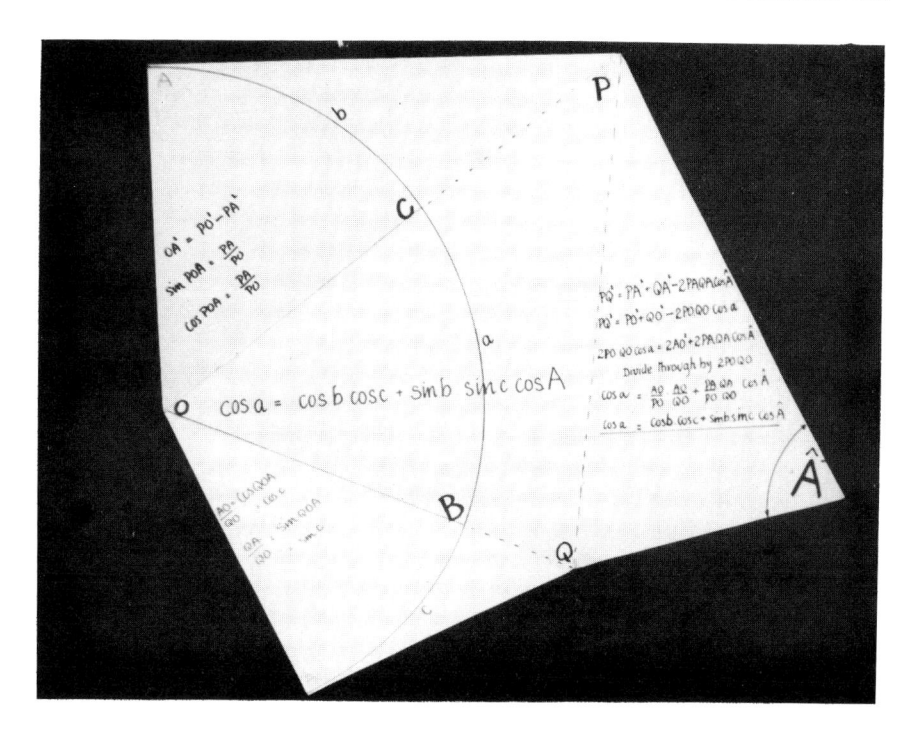

45

Fig. 3.13

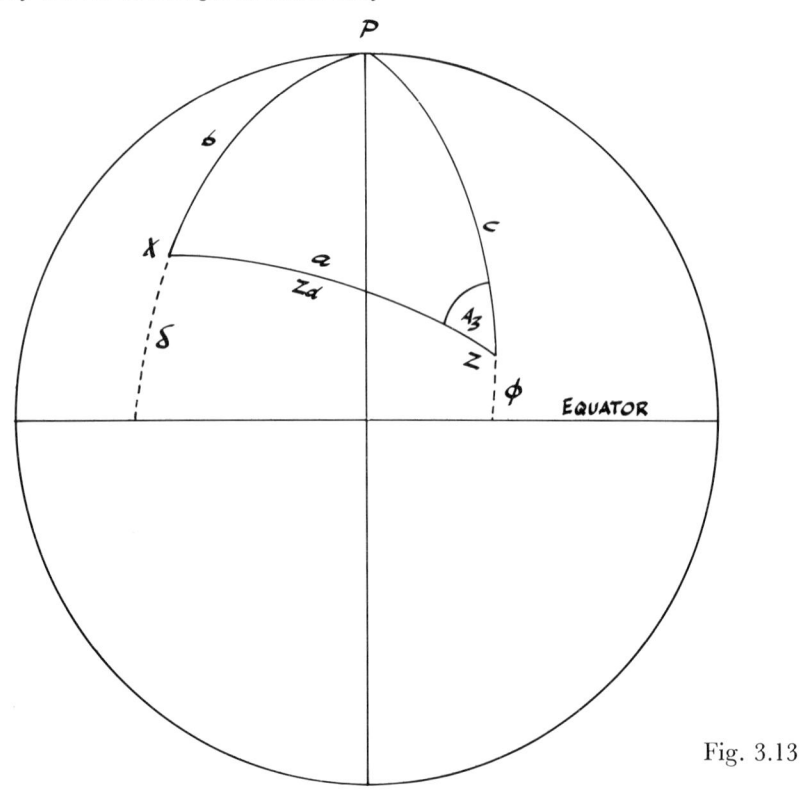

Our fundamental relation becomes,

$$\sin(\text{alt}) = \sin \delta \sin \phi + \cos \delta \cos \phi \cos(\text{HA}) \quad [3.5(1)]$$

This expression is the one most conveniently used in astro-navigation and will be referred to later in 3.7 as *relation* [3.7(1)]. In this era of scientific calculators it should not be regarded as a daunting formula to be worked out using log and trig tables, but as a one-line coded programme. A quick example will suffice to illustrate this point.

Given $\phi = 51°.1$, $\delta = 16°.2$, local hour angle $= 78°.3$

($5^\text{h}\ 13^\text{m}\ 12^\text{s}$), find the altitude of the star.

The calculator programme is:

$$\sin(\text{alt}) = \sin 16.2 \times \sin 51.1 + (\cos 16.2 \times \cos 51.1 \times \cos 78.3)$$

whence altitude $= 19°.8409$. No paper or pencil required, and the solution emerged from the calculator in just 40 seconds! The use of brackets in this expression is necessary when using an algebraic logic calculator (see Section 4.3).

Referring to Fig. 3.13 again, and to relation [3.4], we can turn the spherical triangle round so that our original relation [3.4], becomes

becomes
$$\cos a = \cos b \cos c + \sin b \sin c \cos A$$
$$\cos b = \cos c \cos a + \sin c \sin a \cos B$$

which is the astronomical relation [3.5(2)] below, remembering $b = (90-\delta)$, $c = (90-\text{alt})$ and B is now the azimuth.

$$\sin \delta = \sin(\text{alt}) \sin \phi + \cos(\text{alt}) \cos \phi \cos(\text{Az}) \qquad [3.5(2)]$$

Now we have two powerful means in relations [3.5(1)] and [3.5(2)] for dealing with all the problems involved in switching from equatorial coordinates to alt–azimuth coordinates. If we know the HA and declination we can find the altitude in a few seconds or if we know the azimuth and altitude we can find the declination, knowing the time or hour angle.

In order to help the learner to understand the behaviour of the heavenly sphere as it appears to revolve about the equatorial north–south axis of the Earth the use of a star globe was advocated in Section 2.2.

A star globe can be greatly increased in its usefulness if it is fitted with a simple framework consisting of horizon and altitude circles. These circles can be adjusted in position appropriate to the latitude and can then give approximate values for the altitudes (or depressions) and the azimuths of all objects marked on the globe.

In Fig. 3.14 an example of such a star globe is shown which can be used to illustrate RA, declination, the equinoxes, hour angle, altitude and azimuths, the times of transit, risings and settings.

With this framework of graduated circles all questions of elementary positional astronomy can be answered with an accuracy of a few degrees, so that a helpful rough check on calculated positions can be made.

Such a star globe is invaluable for ensuring that the formulae you are using make sense, and for assigning the right signs when doubt may arise. It helps in star identification and in giving a quick rough check on azimuths or bearings.

3.6 Sight Reduction Tables

Most books on positional astronomy for amateurs or on astro-navigation for seamen used to devote, quite properly, several sections to basic trigonometry and spherical trigonometry and then gave examples which tended to demonstrate how awkward and tedious it was to work them out using trig tables and log tables. Other examples and methods of working intended to facilitate working with logarithms using haversines and half

Positional Astronomy and Astro-Navigation Made Easy

log haversines followed, but these mainly succeeded in demonstrating how much more difficult and incomprehensible the so-called easier methods really were, as they seemed to lose sight of the basic formulae. To make life more tolerable for yachtsmen and airmen and for amateur astronomers, sets of ready reckoner tables have been compiled with tremendous energy and industry which gives the solution of every possible spherical triangle having sides and angles *in whole numbers* likely to be met with. I refer to the six volumes of *Sight Reduction Tables for Air Navigation*.*

Fig. 3.14 A star globe showing the important Great Circles.

*These volumes were compiled by the United States Hydrographic Office with cooperation between the Nautical Almanac Office of the U.S. Naval Observatory, and H.M. Nautical Almanac Office of the U.K.

Spherical Triangles and Early Devices for their Solution

As an indication of the work that has gone into these volumes, there are well over a million entries in some of them, and each entry is a solution to the problem: "Given the latitude, the hour angle and the declination, find the altitude and azimuth of the observed body". (An example of the rapid solution of this type of problem using a calculator has been given in Section 3.4.)

The serious snag with these ready reckoners is that of necessity they have their entries in whole numbers of degrees for HA's, whole numbers for latitudes and whole numbers for declinations. So for any degree of accuracy it is necessary to interpolate, and this can be a tedious and complicated computation despite more sets of tables designed to ease this burden. The navigator using sight reduction tables and interpolation tables is often left with the fear that he may have got into the wrong column, or has got his arthmetic wrong.

3.7 A Table of Useful Relations Used in Positional Astronomy and Astro-Navigation

Practically all problems in positional astronomy concerning hour angles (HA), latitude of observer, ϕ, altitude (alt), declination δ, Right Ascension (RA), and azimuth (Az) can be solved by the scientific electronic calculator using one or more of the following relations of spherical trigonometry.

In addition, Fig. 3.16 gives the basic relations for solving spherical triangles in the form of a family tree with the pride of place being given to *altitude* of a celestial body above the horizon.

[3.7(1)] $\sin(\text{alt}) = \sin\phi \sin\delta + \cos\phi \cos\delta \cos(\text{HA})$

or,

$$\cos(\text{HA}) = \frac{\sin(\text{alt}) - \sin\phi \sin\delta}{\cos\phi \cos\delta}$$

Follows from Pythagoras and plane trigonometry, see models and diagrams in Sections 3.4 and 3.5.

When (HA) = 0 (or 180) we get,

$\sin(\text{alt}) = \sin\delta \sin\phi + \cos\delta \cos\phi = \cos(\phi - \delta)$

∴ (alt) = $90 - \phi + \delta$ hence,

[3.7(2)] $\phi = 90 - (\text{alt}) + \delta$

When (alt) = 0 at rising and setting, then,

[3.7(3)] $\cos(\text{HA}) = -\tan\phi \tan\delta$

This is the basic relation for astro–navigation giving times of rising and setting.

Positional Astronomy and Astro-Navigation Made Easy

[3.7(4)] $\quad \sin \delta = \sin \phi \sin(\text{alt}) + \cos \phi \cos(\text{alt}) \cos(\text{Az})$

or,

[3.7(5)] $\quad \cos(\text{Az}) = \dfrac{\sin \delta - \sin \phi \sin(\text{alt})}{\cos \phi \cos(\text{alt})}$

This is really of the same form as [3.7(1)] with Fig. 3.11 rotated so that (alt) becomes δ and HA becomes (Az).

When (alt) = 0 at rising and setting then

$\sin(\text{alt}) = 0$

and therefore

[3.7(6)] $\quad \cos(\text{Az}) = \dfrac{\sin \delta}{\cos \phi}$ can be used for getting a bearing or checking a compass at sunrise or sunset.

[3.7(7)] $\quad \sin(\text{Az}) = \dfrac{\sin(\text{HA}) \cos \delta}{\cos(\text{alt})}$

used for finding a star's azimuth

See Fig. 3.11. This corresponds to the sine formula of plane trigonometry of Fig. 3.10.

[3.7(8)] $\quad \cot(\text{Az}) = \dfrac{\cos \phi \tan \delta - \sin \phi \cos(\text{HA})}{\sin(\text{HA})}$

This is known as the four part formula, and can be deduced by the ordinary rules of plane trigonometry from [3.7(1), (4) and (7)].

These relations are valid for both the northern and the southern hemispheres, provided values for ϕ in the southern hemisphere, and for δ when the celestial body is south of the celestial equator, are given correct negative values. Hour angles are always positive as they are measured from the meridian in the direction west, north, east, south. Negative values for altitude may be encountered in some calculations, but this means that the value is below the horizon, and is measured towards the nadir. These points are referred to again in Section 5.10.

3.8 The Sun's Hour Angle and its Altitude

Expression [3.7(1)] gives us a means of telling the time by the Sun. We can find the Sun's altitude by a sextant observation or by using one of the simple astrolabic devices described earlier in this chapter. The Sun's declination is given for each day of the year in *Whitaker's Almanack* and if great accuracy be required, for each hour in the *Nautical Almanac*. There are several such publications called ephemerides and they have to be renewed every year.

By rearranging (1) we have,

$$\cos(\text{HA}) = \dfrac{\sin(\text{alt}) - \sin \phi \sin \delta}{\cos \phi \cos \delta}$$

Spherical Triangles and Early Devices for their Solution

which gives us the local hour angle from which local mean time can be found. This is the principle employed in 'altitude sundials' (Chapter 7).

In the case of the Sun,

$$(\text{HA})\,\text{Sun} = \text{GMT} + 12 + E + {}^{+E}_{-W}\,\text{Long}$$

3.9 Times and Azimuths of Risings and Settings of Celestial Bodies Using a Calculator

From [3.7(1)] the altitude at rising and setting $= 0$, so

$$0 = \sin\phi\,\sin\delta + \cos\phi\,\cos\delta\,\cos(\text{HA}),$$

or

$$\sin\phi\,\sin\delta = -\cos\phi\,\cos\delta\,\cos(\text{HA}),$$

and

$$\cos(\text{HA}) = -\tan\phi \times \tan\delta.$$

From this relation it is easy to calculate the hour angle of a body at rising or setting. For example, let us take an annual event of particular interest, namely sunrise and sunset at Stonehenge, lat $51°.17$ on 21st June. The declination of the Sun $= 23°.44$.

$$\cos(\text{HA}) = -\tan 51.17 \times \tan 23.44$$
$$(\text{HA}) = 122°.59$$
$$= 8.1729\,\text{hr}.$$

Approximately therefore, neglecting the refraction of the air and taking the centre of the Sun on the horizon as marking sunrise or sunset, we have sunrise $8^h\,10.37$ min before local noon, i.e., $3^h\,49.23^m$ a.m. local true sun time, and sunset $8^h\,10.37$ min after local noon, or $20^h\,10.37^m$ local true sun time.

These times have to be corrected to give UT by adding $7^m\,20^s$ (because the longitude of Stonehenge is $1°\,50'$ W) and then subtracting $1^m\,44^s$ for the equation of time corresponding to 21st June. The refraction of the atmosphere is considered in Sections 3.24 and 4.7(2).

3.10 Amplitudes—Bearings of the Sun when Rising or Setting

Just *where* in *azimuth* the Sun rises or sets at these times is another simple one finger exercise on the calculator using equation [3.7(4)], namely,

$$\sin\delta = \sin\phi\,\sin(\text{alt}) + \cos\phi\,\cos(\text{alt})\,\cos(\text{Az}).$$

We can readily calculate the azimuth of the rising and setting of the Sun. This is often a great help to navigators in obtaining the compass error.

Positional Astronomy and Astro-Navigation Made Easy

At rising and setting (alt) = 0 and sin(alt) = 0 and cos(alt) = 1. Whence, from [3.7(6)]

$$\cos(Az) = \frac{\sin \delta}{\cos \phi}$$

which is independent of time or hour angle. The amplitude is the only astronomical sight that requires no instrument as it depends only on the Sun's declination and the observer's latitude. Nature provides the horizon circle.

For example, midsummer sunrise at Stonehenge:

$$\cos(Az) = \frac{\sin 23.44}{\cos 51.17} \qquad (Az) = 50°.623$$

For sunset, Azimuth is $360 - 50.623 = 309°.377$

At the midwinter solstice the calculator takes care of the change in sign of the Sun's declination, $-23°.44$, and gives azimuth for rising,

$$\cos(Az) = -\frac{\sin 23.44}{\cos 51.17} \qquad (Az) = 129°.377,$$

and the azimuth at the Sun's setting $= 360° - 129°.377 = 230°.632$.

This simple formula applied to the various extremes of the Moon's declinations over the centuries has helped to establish the belief that Stonehenge was a lunar observatory some 2000 years ago (see Section 3.15).

It is an interesting project for the use of a calculator to tabulate values of hour angles and declinations using [3.7(3)] and on the same graph paper tabulate values of azimuths and declinations using relation [3.7(6)]. The two resulting curves from plotting the points are shown in Fig. 3.15 (neglecting atmospheric refraction) from which it is noted that azimuths and hour angles are supplementary when declinations are 0 and when they have values $\pm(90-\phi)$ on the southern and northern hemispheres respectively.

In Fig. 3.15 the continuous curve shows the relation between the times of rising and setting of celestial bodies and their declinations. For stars the abscissa is the hour angle in degrees. For the sun the times can be expressed in hours, as the local sun time + 12. The relation, (neglecting refraction) is $\cos HA = -\tan \phi \tan \delta$.

The broken curve shows how the azimuths of bodies vary with bodies declinations on rising and setting. The azimuths being measured from the North point on the horizon towards the East. The relation in this case is

$$\cos(Az) = \frac{\sin \delta}{\cos \phi} \qquad \text{Latitude } \phi = 51°$$

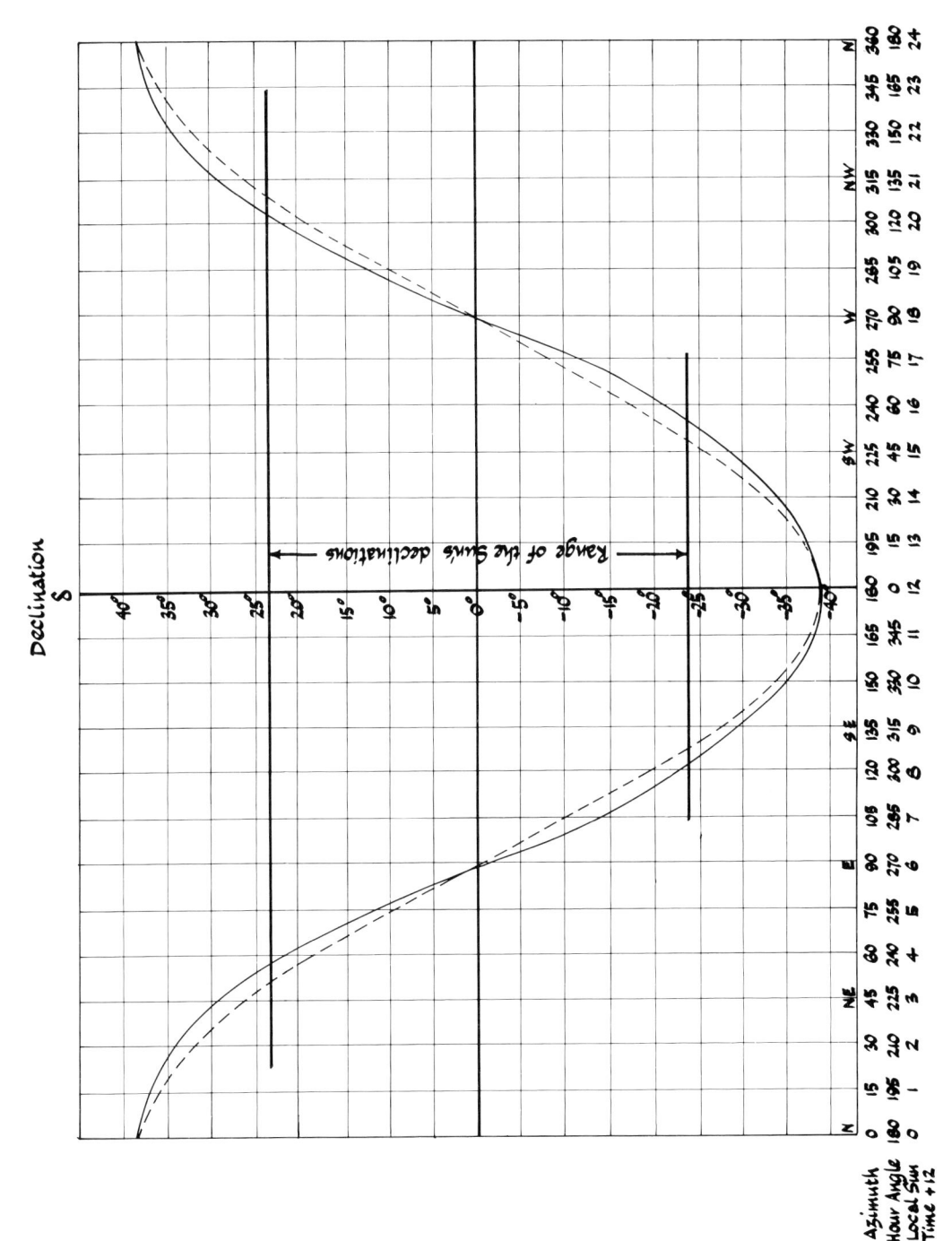

Spherical Triangles and Early Devices for their Solution

Fig. 3.15

3.11 The Prime Vertical Altitude

When a celestial body is due east or due west, it is said to be on the 'prime vertical', with (Az) = 90° or 270° so then [3.7(4)] becomes

$$\sin(\text{alt}) = \frac{\sin \delta}{\sin \phi}$$

This relation used in conjunction with a pre-set sextant and a calculator can determine due east or west and so provides a means of finding the compass error (see Section 4.22).

3.12 The Four Part Formula

The relation [3.7(8)] is often known as the 'four part' formula as it involves four *consecutive* parts of a spherical triangle, and can be deduced by the ordinary rules of trigonometry from [3.7(1)], [3.7(4)] and [3.7(7)]. The derivation is given in Appendix I. It will be used in Chapter 5 in dealing with alt–azimuth curves. It is not easy to remember or to apply but the formula can be made a little more helpful if we give the four consecutive angles and sides the numbers (1), (2), (3) and (4). Then:

$$\cos(2)\cos(3) = \sin(2)\cot(4) - \sin(3)\cot(1)$$

Using these relations in Section 3.7, and the scientific calculator, we now have at our fingertips the answers to a wide range of problems in positional astronomy and astro-navigation. Special reference is made in Chapter 7 to various kinds of sundials and the relations that connect 'shadow angles' with hour angles and latitude.

3.13 Local Hour Angles

We note that the LHA occurs in relations [3.7(1), (3) and (4)] and, as this changes with time, it must be found for a precise instant in sidereal or mean time.

In finding positions of heavenly bodies, or your own position, the calculator can be used to do the simple arithmetic necessary for finding the hour angles of celestial bodies and for reducing hours, minutes and seconds to degrees, minutes and seconds of arc, or to degrees and decimals.

Many calculators with trigonometrical functions deal only with angles expressed in degrees and decimal fractions of a degree, so that an angle expressed in degrees, minutes and seconds of arc has to be changed. This is very easily done. For example, to express 29°37′21″ in degrees and a

Spherical Triangles and Early Devices for their Solution

decimal, enter 21" and divide by 60. Display = 0.35. Add 37′ read 37.35, divide this by 60. Display = 0.622 5.

$$\therefore \quad 29°37'21'' = 29°.622\ 5.$$

No paperwork is required. The reverse process will readily suggest itself.

Some of the more expensive calculators have special keys for these transformations.

A useful relation for finding the local hour angle of a star is given in Section 2.7 and, as we have seen, we can calculate the LHA if we know the ST at 0^h and GMT, bearing in mind that we have to increase GMT by approximately 10 seconds every hour to bring it in line with sidereal time. Thus, if T is the mean time in hours, then approximately:

$$(LHA) = (ST\ at\ 0^h) + T + 10T\ secs\ {}^{+E}_{-W}\ Long - (RA)$$

or, more accurately as we have seen, from Section 2.9

$$(LHA) = ((ST\ at\ 0^h) + 1.002\ 738\ T){}^{+E}_{-W}\ Long - (RA). \quad [3.13(1)]$$

The part in brackets is simply the local sidereal time, and we have as in Section 2.7

$$(LHA)* = local\ sidereal\ time - (RA)*. \quad [3.13(2)]$$

For the Sun, the LHA = GMT + 12 + $E {}^{+E}_{-W}$ Long, where E is the equation of time, as mentioned in Section 2.7.

To find the LHA of a star, the relation [3.13(1)] above can be used, as follows:

From *Whitaker's Almanack* the sidereal time at 0^h can be found for the day and the star's RA is given in the astronomical section. For navigational purposes and more accurate work we use the relation [3.13(2)] and we know that

$$local\ sidereal\ time = local\ hour\ angle\ of\ \Upsilon$$
$$= (GHA)\ {}^{+E}_{-W}\ Long$$

As mentioned in Section 2.8, the *Nautical Almanac* does not give the RA's of stars but prefers to measure the angular gap between Υ and the star by measuring in the direction of the apparent motion of stars. This measure is called the star's 'sidereal hour angle', or SHA, which is 360 − (RA) and as 360° makes no difference in an hour angle problem we can write,

$$(LHA)* = (GHA)\ \Upsilon\ ({}^{+E}_{-W}\ Long) + (SHA)*$$

3.14 The Importance of Altitude

A diagram showing *altitude* at the head of a family of mathematical relations used in astronomy is shown in Fig. 3.16.

ALTITUDE

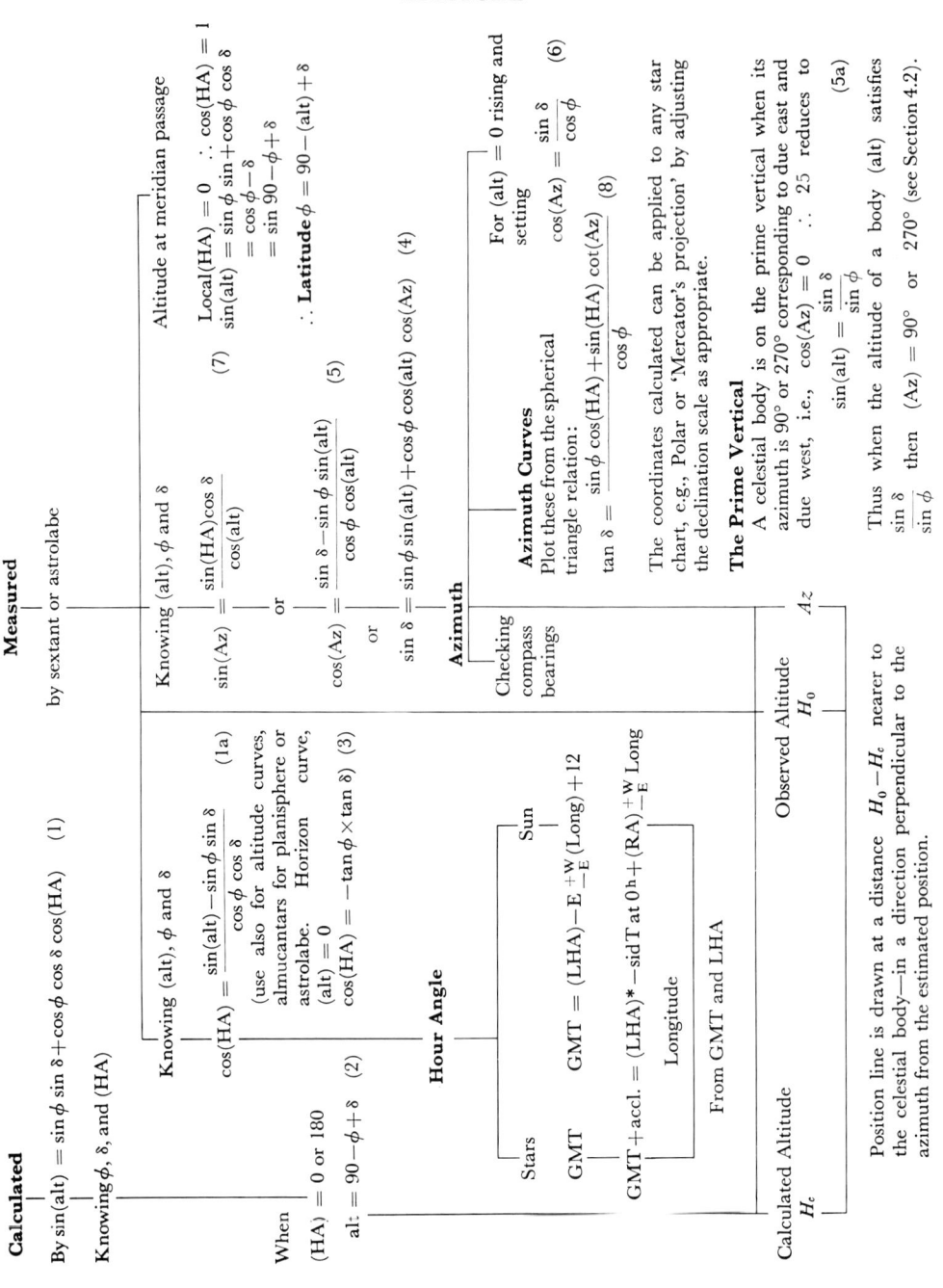

Calculated

By $\sin(\text{alt}) = \sin\phi \sin\delta + \cos\phi \cos\delta \cos(\text{HA})$ (1)

Knowing ϕ, δ, and (HA)

Measured

by sextant or astrolabe

Knowing (alt), ϕ and δ

$$\cos(\text{HA}) = \frac{\sin(\text{alt}) - \sin\phi \sin\delta}{\cos\phi \cos\delta} \quad (1a)$$

(use also for altitude curves, almucantars for planisphere or astrolabe. Horizon curve,

When
(HA) = 0 or 180
al: = $90 - \phi + \delta$ (2)

$\cos(\text{HA}) = -\tan\phi \times \tan\delta$ (3)

Altitude at meridian passage

$\text{Local(HA)} = 0 \quad \therefore \cos(\text{HA}) = 1$
$\sin(\text{alt}) = \sin\phi \sin + \cos\phi \cos\delta$
$= \cos\phi - \delta$
$= \sin 90 - \phi + \delta$ (7)

$\therefore$ **Latitude** $\phi = 90 - (\text{alt}) + \delta$

Knowing (alt), ϕ and δ

$$\sin(\text{Az}) = \frac{\sin(\text{HA})\cos\delta}{\cos(\text{alt})}$$

or

$$\cos(\text{Az}) = \frac{\sin\delta - \sin\phi \sin(\text{alt})}{\cos\phi \cos(\text{alt})} \quad (5)$$

or

$\sin\delta = \sin\phi \sin(\text{alt}) + \cos\phi \cos(\text{alt}) \cos(\text{Az})$ (4)

Azimuth

Checking compass bearings

Azimuth Curves

Plot these from the spherical triangle relation:

$$\tan\delta = \frac{\sin\phi \cos(\text{HA}) + \sin(\text{HA}) \cot(\text{Az})}{\cos\phi} \quad (8)$$

The coordinates calculated can be applied to any star chart, e.g., Polar or 'Mercator's projection' by adjusting the declination scale as appropriate.

The Prime Vertical

A celestial body is on the prime vertical when its azimuth is 90° or 270° corresponding to due east and due west, i.e., $\cos(\text{Az}) = 0 \quad \therefore 25$ reduces to

For (alt) = 0 rising and setting $\cos(\text{Az}) = \frac{\sin\delta}{\cos\phi}$ (6)

$$\sin(\text{alt}) = \frac{\sin\delta}{\sin\phi}$$ (5a)

Thus when the altitude of a body (alt) satisfies $\frac{\sin\delta}{\sin\phi}$ then (Az) = 90° or 270° (see Section 4.2).

Hour Angle

Stars ─── Sun

GMT   GMT = (LHA) $-\text{E}_{-\text{E}}^{+\text{W}}$ (Long) $+12$

GMT + accl. = (LHA)* $-$ sidT at 0h $+$ (RA) $_{-\text{E}}^{+\text{W}}$ Long

Longitude

From GMT and LHA

Calculated Altitude H_c Observed Altitude H_o Az

$H_o - H_c$ nearer to

Position line is drawn at a distance the celestial body—in a direction perpendicular to the azimuth from the estimated position.

Fig. 3.16

56

Spherical Triangles and Early Devices for their Solution

In Section 3.1 we saw that the altitude of a celestial body was about the only simple direct measurement we can usefully make, and in Section 3.2 were described, as suggested projects, five simple devices for 'taking the heights' of celestial bodies.

Now that we have been introduced to the spherical triangle and the various relations between its sides and angles, it is now worthwhile making a diagram showing altitude as a natural and historical starting point for the development of positional astronomy. The diagram also shows how altitude is associated with azimuth and with HA, ϕ and δ in this development.

The diagram is self-explanatory and is intended to summarise the relationships already considered, under Section 3.7 and which will be used in the chapters to follow.

3.15 Stonehenge

A note on Stonehenge may suggest an exercise using the calculator. The monument is a record of activities of many generations spread over 4000 years, and the people achieved astonishing accuracy in solar and lunar alignments using pegs and lines and basic mathematical capabilities including the use of the properties of the right angled triangle. All civilizations have been concerned with the making and revising of calendars based on the phases of the Moon and the annual motions of the Sun.

Lunar investigations appear to have been the concern of many of the ancient megalithic sites in Britain and in Brittany but there is no evidence that Stonehenge was used for alignments on stars. This is probably because the positions of stars change considerably in the course of a few centuries on account of the precession of the Earth's axis, which precesses through 360° every 26 000 years (see Chapter 9). If we were to visit the Earth in the year A.D. 15 000 our Pole Star would be Vega. During the precession the Earth's axis maintains a fairly constant inclination of about 23°.5 to the plane of the ecliptic and changes about 0°.01 every century. This should not be confused with the precessional change in the direction of the Earth's axis of about 50" each year. At present the inclination or obliquity of the ecliptic is 23°.44 and this has changed only about half a degree during the past 4000 years. The inclination of the Earth's axis to the plane of the ecliptic was about 24°, 4000 years ago. Astronomers and archaeologists make use of this small change to help in their dating of ancient *lunar* and *solar* astronomical monuments.

3.16

Stonehenge is situated at a spot with a good horizon all round (an almost perfect azimuth circle) on which could be marked, by means of distant

Positional Astronomy and Astro-Navigation Made Easy

backsights, the precise points of the maximum and minimum full Moon risings and settings, at the summer and winter solstices or near them.

The calculator can tell us very easily just where these risings and settings took place as viewed from the centre of Stonehenge. The azimuth of a rising or setting heavenly body is given by,

$$\cos(\text{Az}) = \frac{\sin \delta}{\cos \phi}$$

The Sun's declination 4000 years ago was $24°$ in mid-summer and $-24°$ in mid-winter.

The Moon has extreme declinations of $24° \pm 5°$ and $-24° \pm 5°$ because the Moon's orbit is inclined $5°$ to the ecliptic and its declination changes from $24+5$ to $24-5$ and back to $24+5$ in a period of about 18 years. The azimuth of the full rising Moon at Stonehenge in mid-winter, neglecting the effects of atmospheric refraction azimuth is given by,

$$\cos(\text{Az}) = \frac{\sin 29}{\cos 51.17} \qquad (\text{Az}) = 39°.35.$$

The minimum mid-winter moonrise (nine years later) is given by,

$$\cos(\text{Az}) = \frac{\sin 19}{\cos 51.17} \qquad (\text{Az}) = 58°.72$$

similarly the minimum mid-summer moonrise is given by,

$$\cos(\text{Az}) = \frac{\sin -19}{\cos 51.17} \qquad (\text{Az}) = 121°.28.$$

Maximum mid-summer moonrise is given by,

$$\cos(\text{Az}) = \frac{\sin - 29}{\cos 51.17} \qquad (\text{Az}) = 140°.64.$$

There are four corresponding directions for the *setting* full Moon in its extreme positions using the same formula. The calculator takes good care of the signs if programmed correctly. The setting positions are found from the fact that cos(Az) is the same for $39°.35$ as for $(360-39.35)$ or $320°.25$ and this is usually written N39.35W.

Five identifiable mounds or marks on the distant horizon several miles from Stonehenge have been found in the directions we have calculated. A full account of this work appeared in the *Journal for the History of Astronomy*, February 1975, under the title 'Stonehenge as a Lunar Observatory' by Alexander Thom.

An interesting feature of these azimuth calculations is the fact that the azimuth line for the Moon's most southerly declination $-29°$ (140.65) is practically at right angles to the Sun's azimuth at the Sun's maximum declination at the summer solstice (49°.53). This fact makes for an interesting rectangular structure involving the four station stones. This situation could arise only at the latitude of Stonehenge.

3.17 The Four Main Coordinate Systems Used in Positional Astronomy

It has been mentioned that when we wish to tell someone where to look for a star or planet on a particular date at a particular time we find it convenient and rational to refer to the star coordinate system we know so well in our right angle dominated space. We usually give the star's altitude and azimuth. This is a consequence of our daily experience and our keen sense of the vertical and horizontal which, if we are not born with, we soon acquire through our early experiences with gravity! We have also seen that when we consider the celestial sphere as apparently turning about us on an axis through the N and S poles of the heavens, and therefore parallel to our own Earth's axis, we define a star's position by its Right Ascension and its declination in a manner analogous to defining a geographical position by longitude and latitude on the Earth's sphere.

For our practical purposes in positional astronomy, we consider the origin of our coordinate system as being at the centre of the Earth and we recognise four main reference systems for defining positions.

System	*Positions determined by:*
(1) The Horizon and Local Meridian.	Altitude and Azimuth (Fig. 3.17a).
(2) The Celestial Equator and the Equinox.	Declination and Right Ascension. These coordinates fix the position of a body on the celestial sphere and are independent of daily time (Equatorial system) (Fig. 3.17b).
(3) The Celestial Equator and the Local Meridian.	Declination and Hour Angle. These coordinates are dependent on time (Equatorial system) (Fig. 3.17b).
(4) The Ecliptic and the Equinox.	Celestial Latitude and Longitude (this system is considered later in Chapter 9).

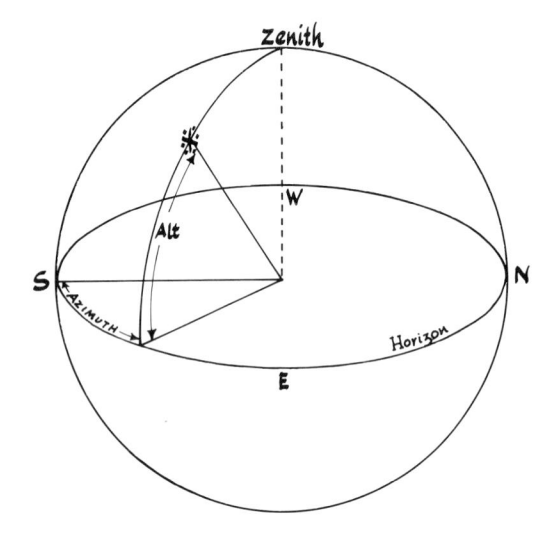

Alt-Azimuth system of co-ordinates

Fig. 3.17(b)

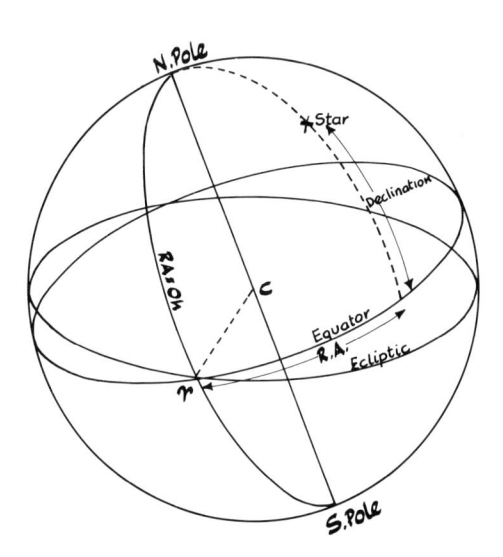

Equatorial or R.A. declination system

Fig. 3.17(a)

Spherical Triangles and Early Devices for their Solution

3.18 An Alt–Azimuth Model

A simple alt–azimuth instrument can be constructed from two protractors, one semi-circular, for measuring altitudes and the other a fixed circular protractor graduated from 0° to 360° for measuring azimuths (see Fig. 3.18).

This azimuth circle is kept horizontal and with the 0°–180° marks lying N–S. The diagram shows the principle on which it works and the photograph shows some detail to enable you to make the device using a spirit level which gives the vertical line, which serves as a plumbline. The model makes use of small nuts and bolts and perspex which is strong, transparent and easily cut and drilled (see Figs. 3.18 and 3.19).

3.19

A Home-made Equatorial Device which connects systems (2) and (3) of Section 3.17 and shows the relationships between declination, Right Ascension and sidereal time.

This is best described by reference to Fig. 3.20. Instead of an azimuth circle of Fig. 3.19, a RA circle showing the main stars of the northern hemisphere (or southern hemisphere) is used showing hour angles from 0^h to 24^h or 0° to 360° (see Fig. 3.20).

Positional Astronomy and Astro-Navigation Made Easy

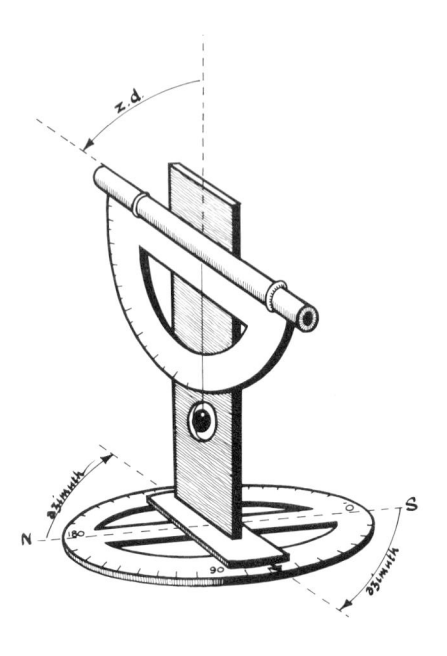

Fig. 3.19. Home-made apparatus for measuring azimuth as well as altitude of a star.

This RA circle rotates about its centre, the North Pole, in a clockwise direction as viewed along the polar axis from north to south. The RA circle used in the photograph is a common type of star map with polar coordinates, favoured by navigators for star identification, and formerly published by the Hydrographic Office, Washington D.C., under the authority of the U.S. Navy.

It is important to note how this star map differs from the polar star map of the planisphere described in Section 2.4 which rotates in an anti-clockwise direction because the observer is considered to view the RA circle along the polar axis from the South Pole towards the North Pole. Confusion can arise unless this difference be appreciated.

The plane of the RA circle is parallel to the equator and therefore inclined to the horizontal at an angle of $90 - \phi$ (where ϕ is the observer's latitude). The axis about which the 180° protractor turns is parallel to the Earth's polar axis. The protractor now measures the declination of a star which is observed along the sighting bar of the protractor.

Spherical Triangles and Early Devices for their Solution

Telescope users will recognise in this model the essential features of an equatorial telescope fitted with setting circles, and the model may help beginners to use setting circles and sidereal time with full understanding and skill. The sighting bar represents the telescope, the degree marks on the 180° protractor represents the declination setting circle. The simulated telescope turns about the polar axis and the RA circle represents the RA setting circle.

To bring the simulated 'telescope' on to a particular star:

(1) Turn the protractor sighting bar and the pointer, C, on to the RA of the star to be observed. This is equivalent to putting the telescope on to the appropriate RA on the setting circle for the star.

(2) Now turn the RA disc together with the telescope so that the local sidereal time (from a nautical almanac) is on the meridian line of the base (S in Fig. 3.22).

(3) Turn the declination protractor on to the star's declination. The sighting bar or telescope should now have the star in its field of view.

The RA setting circle is used in this way to bring the telescope on to the star's hour angle as it effects the subtraction sum

$$(\text{LHA}) = \text{local sidereal time} - (\text{RA}).$$

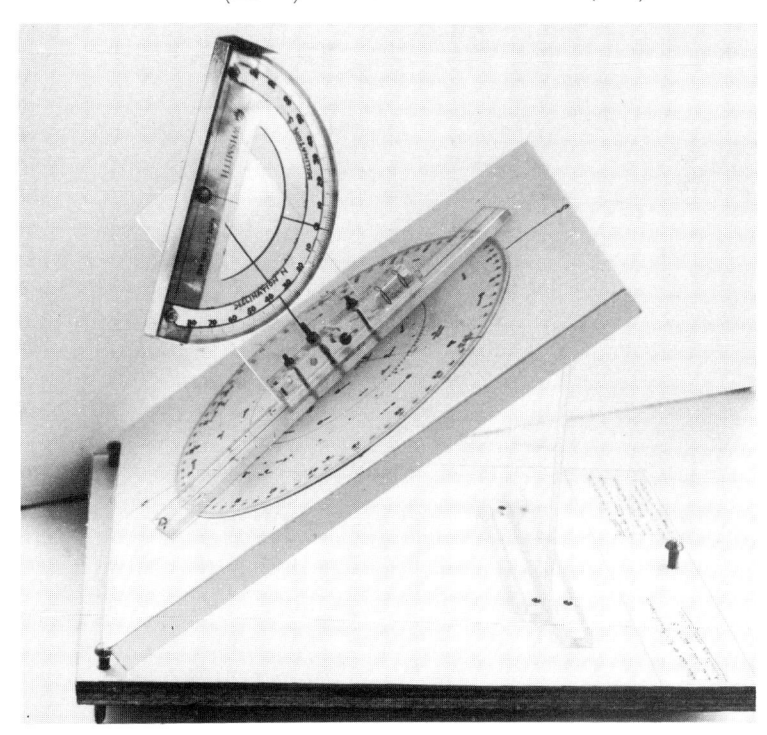

Fig. 3.20

Positional Astronomy and Astro-Navigation Made Easy

Example: (See Figs. 3.20, 3.21 and 3.22 for the same situation as shown in Fig. 3.23 on a telescope with setting circles)

$$\text{local sidereal time} = 3 \text{ hr.}$$
$$(RA) = 4^h 30^m.$$
$$\begin{aligned}\text{local hour angle} &= \text{sidereal time} - (RA) \\ &= 3^h - 4^h\, 30^m + 24^h \\ &= 22^h\, 30^m\end{aligned}$$

3.20 It will be noticed that by tilting an alt–azimuth instrument through $(90-\phi)°$ about an east–west axis we can obtain the basic mechanism for an equatorial mounting.

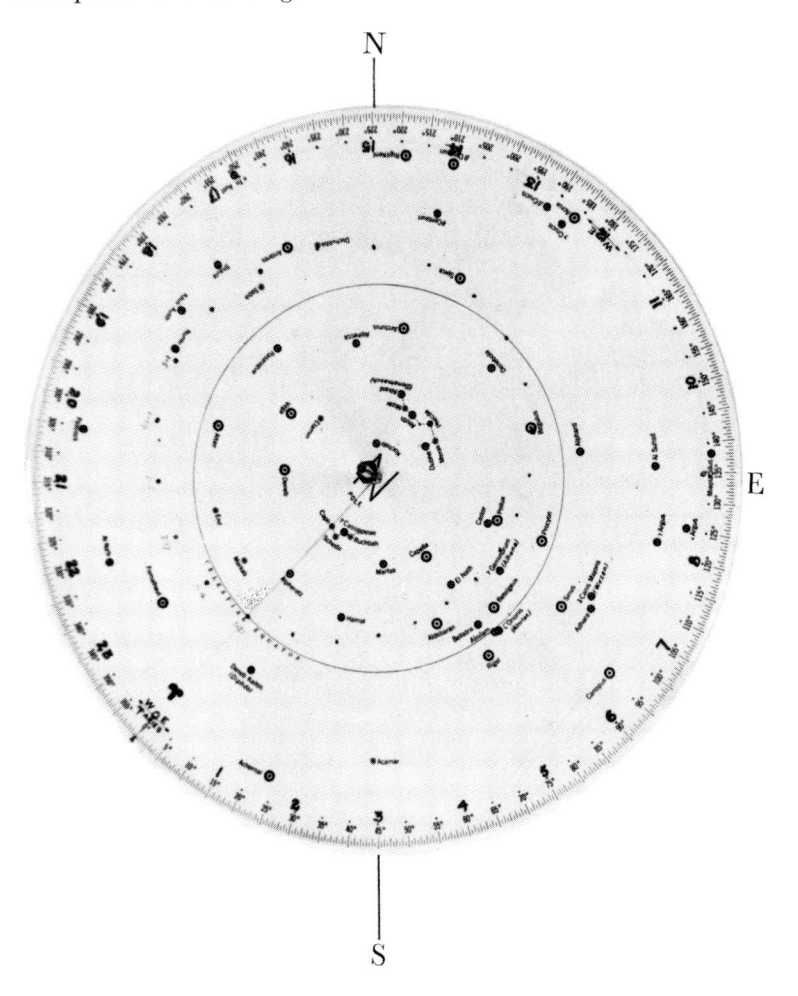

Fig. 3.21

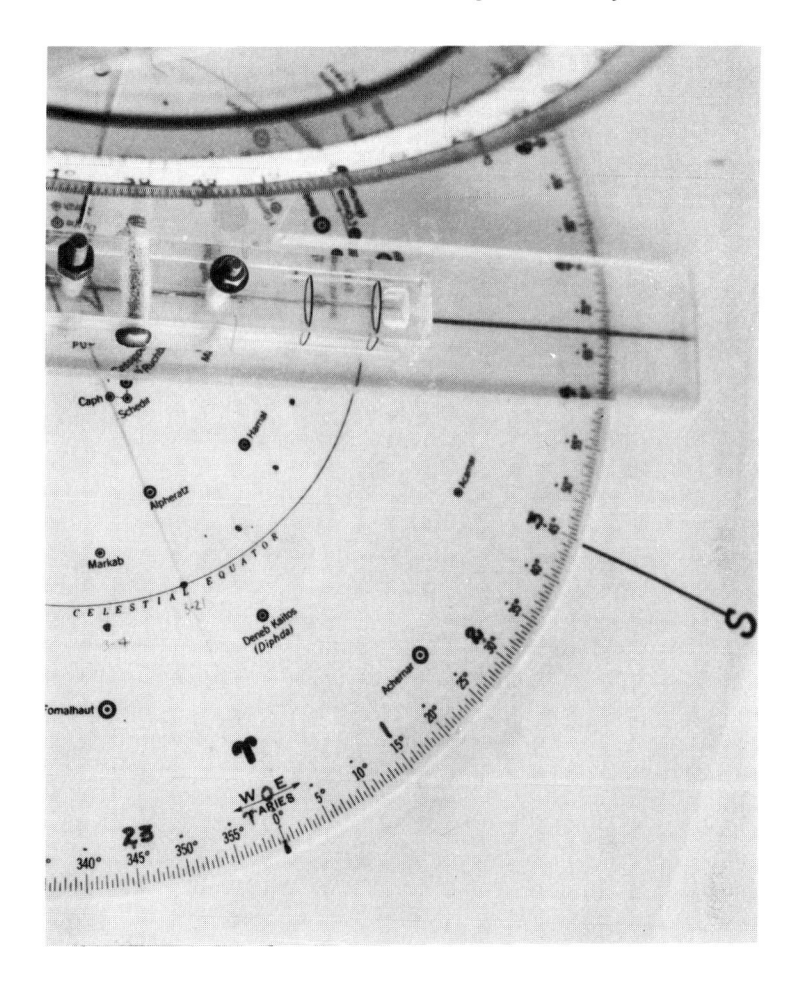

Spherical Triangles and Early Devices for their Solution

Fig. 3.22

Fundamental relations [3.7(1)] and [3.7(7)] of spherical trigonometry can be checked practically using the devices described above using any distant object, and can be verified theoretically using the relations:

$$\sin(\text{alt}) = \sin\phi \sin\delta + \cos\phi \cos\delta \cos(\text{LHA}) \qquad [3.7(1)]$$

$$\sin\delta = \sin\phi \sin(\text{alt}) + \cos\phi \cos(\text{alt}) \cos(\text{Az}) \qquad [3.7(4)]$$

$$\sin(\text{Az}) = \frac{\sin(\text{LHA}) \cos\delta}{\cos(\text{alt})} \qquad [3.7(7)]$$

3.21 The usefulness of these relations will be appreciated from the following examples.

A star is observed to have an altitude of 39°56′ (39°.933) and an azimuth of 44°12′ (44°.2) from the 180° bearing or 224°.2 true bearing, when the local sidereal clock shows time 2200 hr. Time is an essential connecting link between the two sets of coordinates. What is the star's declination, its local hour angle and its RA? What is the star?

This is a matter for switching a star's alt–az coordinates to RA–declination coordinates. Apply relation [3.7(4)] to find the star's declination from altitude and azimuth, using,

$$\sin \delta = \sin \phi \sin(\text{alt}) + \cos \phi \cos(\text{alt}) \cos(\text{Az})$$

i.e. $\sin \delta = \sin 51° \sin 39°.933 + \cos 51° \cos 39°.933 \cos 224°.2$

whence $\delta = 8°.794.$

We can now use the declination in

$$\sin(\text{LHA}) = \frac{\sin(\text{Az}) \cos(\text{alt})}{\cos \delta}$$

$$= \frac{\sin 44.2 \cos 39.933}{\cos 8.794}$$

whence $(\text{LHA}) = 32°.748.$

Also, $(\text{LHA}) = (\text{ST}) - (\text{RA})$

$32.748 = 330 - (\text{RA})$

$(\text{RA}) = 330 - 32.748$

$= 297.252$

$= 19^\text{h}.8168$ or $19^\text{h}\ 49^\text{m}.$

The star, therefore, has declination 8.794°N and RA $19^\text{h}\ 49^\text{m}$ and can be identified as Altair.

3.22 If the RA and the declination of a star be known, it can readily be identified from a good star map or almanac. Stars can also be identified in terms of RA and declination from two simple basic observations, altitude and azimuth, provided we know or can find the local sidereal time.

The switching of coordinates can of course be done from RA–declination coordinates to alt–azimuth coordinates by using the same relations, as in the following example.

Consider Procyon at midnight on 20th February.

$$(\text{RA}) = 7^\text{h}\ 38.1^\text{m}$$

$$\delta = 5°17'\ (5°.283).$$

Given, local sidereal time $= 10^\text{h}\ 0^\text{m}.$

Spherical Triangles and Early Devices for their Solution

Fig. 3.23. Actual telescope showing sidereal time, 3^h. RA of star, $4^h\ 30^m$.
Local hour angle $= 3^h - 4^h\ 30^m + 24^h$
$= 22^h\ 30^m$.

The star's local hour angle $=$ sidereal time $-(\text{RA})$
$= 10^h\ 0^m - 7^h 38.1^m$
$= 2^h\ 21.9^m$
$= 35°.475$.

The altitude, using relation [3.7(1)], is found from

$\sin(\text{alt}) = \sin 51 \sin 5°.283 + \cos 51° \cos 5°.283 \cos 35.475$

altitude $= 35°.582$.

Using relation [3.7(7)] for azimuth,

$$\sin(\text{Az}) = \sin \frac{35.475 \cos 5°.283}{\cos 35.582}$$

$(\text{Az}) = 45°.28$ from the $180°$ mark, or the *bearing* $= 180 + 45.28$
$= 225°.28$.

We can therefore find Procyon by looking in a direction bearing $225°.28$ azimuth and $35°.582$ altitude from the horizon.

3.23 Angles of Stars as they Rise and Set, with Respect to the Horizon

An interesting little problem was suggested as a calculator exercise by the rising and setting phenomenon of the Sun or Moon at Stonehenge, namely: 'What angle does a rising body make with the horizon, given its declination and the latitude of the place?' (See Fig. 3.24.)

This angle was required in order to decide at which point on the horizon the Sun rose, or how much of the Sun had to be above the horizon before it could be said that it had risen, e.g., the top tip, the centre, or the disc quite clear. The Sun's disc is about $\frac{1}{2}°$ in diameter so the angle of rising affects the azimuth and the hour angle of its rising.

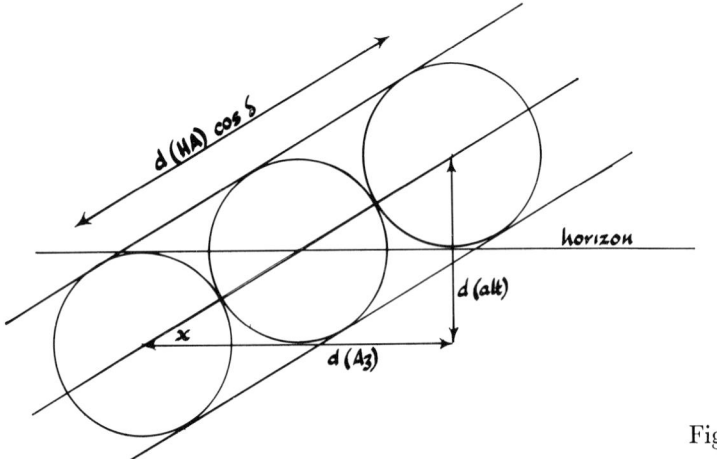

Fig. 3.24

We can use relation (1):

$$\sin(\text{alt}) = \sin \phi \sin \delta + \cos \phi \cos \delta \cos(\text{HA}). \qquad [3.23(1)]$$

We require the angle x in terms of ϕ and δ, e.g.,

$$\sin x = \frac{d(\text{alt})}{d(\text{HA}) \cos \delta}. \qquad [3.23(2)]$$

Differentiate (1):

$$\cos(\text{alt}) \, d(\text{alt}) = \cos \phi \cos \delta \sin(\text{HA}) \, d(\text{HA}) \qquad [3.23(3)]$$

where ϕ and δ are constants.

From [3.23(2)] and [3.23(3)]

$$\sin x = \frac{d(\text{alt})}{d(\text{HA}) \cos \delta} = \frac{\cos \phi \sin(\text{HA})}{\cos(\text{alt})}$$

Spherical Triangles and Early Devices for their Solution

By the sine relation [3.7(7)]

$$\frac{\sin(\text{HA})}{\cos(\text{alt})} = \frac{\sin(\text{Az})}{\cos \delta}.$$

Therefore
$$\sin x = \frac{\cos \phi \sin(\text{Az})}{\cos \delta}. \qquad [3.23(4)]$$

From [3.7(6)] we know that

$$\cos(\text{Az}) = \frac{\sin \delta}{\cos \phi} \quad \text{when the (alt)} = 0 \qquad [3.23(5)]$$

so we can eliminate Az from [3.23(4)] and [3.23(5)] and so obtain an expression for x in terms of ϕ and δ.

From [3.23(4)]
$$\sin(\text{Az}) = \frac{\sin x \cos \delta}{\cos \phi}$$

and from [3.23(5)]
$$\cos(\text{Az}) = \frac{\sin \delta}{\cos \phi}.$$

By squaring and adding [3.23(4)] and [3.23(5)] we have

$$\sin^2(\text{Az}) + \cos^2(\text{Az}) = 1$$

and therefore,

$$\frac{\sin^2 x \cos^2 \delta}{\cos^2 \phi} + \frac{\sin^2 \delta}{\cos^2 \phi} = 1$$

$$\therefore \quad \cos^2 \phi = \sin^2 x \cos^2 \delta + \sin^2 \delta$$

$$\sin^2 x = \frac{\cos^2 \phi - \sin^2 \delta}{\cos^2 \delta}$$

$$= \frac{\cos^2 \phi - (1 - \cos^2 \delta)}{\cos^2 \delta}$$

$$= \frac{\cos^2 \phi - 1 + 1}{\cos^2 \delta}$$

$$\therefore \quad 1 - \sin^2 x = \frac{\cos^2 \phi}{\cos^2 \delta}$$

$$\cos x = \frac{\sin \phi}{\cos \delta}$$

so the angle of rising or setting of a body depends only on ϕ and δ.

For the Sun at Stonehenge at midsummer in 2000 B.C.,

$$\cos x = \frac{\sin 51.1}{\cos 24} \qquad x = 31°.58 = \text{the angle of rising and setting.}$$

Positional Astronomy and Astro-Navigation Made Easy

It is worthy of note that this angle is the same for $+\delta$ as for $-\delta$ and when the declination $\delta = 0$, x is a *maximum* and is equal to the *complement of the latitude*, and when the declination is equal to the complement of the latitude (i.e. $+38°.9$ in the example) then $x = 0$ and the star just touches the horizon. These calculations do not take into account the refraction of the Earth's atmosphere, which at rising and setting slightly affects the result.

Risings and settings are steepest for bodies on the celestial equator but it is interesting and useful at times to know just what angle a star or planet will make with the horizon on rising or setting, but it should not be expected that the path will be followed accurately for more than a few degrees, as the path is part of a circle described on the celestial sphere.

3.24 Effect of Refraction on Times of Rising and Setting

The values calculated in Section 3.16 are approximate as we have ignored the effect of the refraction of the atmosphere, which is regarded for practical purposes, as a constant, 34' for heavenly bodies at zero altitude.

This effect of refraction is important in calculating the exact times of sunrise and sunset or in calculating the length of the day at a particular place on a particular date.

We can make these calculations readily as the following example will show.

(1) To find the effect of refraction on the length of the day, that is the interval between the first appearance and the disappearance of the Sun's upper limb at a latitude of 51°N declination 13°N see Fig. 3.26.

We use the relation [3.7(1)],

$$\sin(\text{alt}) = \sin\phi \sin\delta + \cos\phi \cos\delta \cos(\text{HA})$$

which, when the altitude of the Sun is theoretically zero (i.e., when its centre is on the horizon and there is no refraction) (Fig. 3.26) it becomes

$$0 = \sin\phi \sin\delta + \cos\phi \cos\delta \cos(\text{HA})$$

and

$$\cos(\text{HA}) = -\tan\phi \tan\delta$$
$$\cos(\text{HA}) = -\tan 51 \tan 13$$
$$(\text{HA}) = 106°.57 \quad (\text{or } 286°.57).$$

Now find the hour angle of the Sun when its upper limb first appears (Fig. 3.25), i.e., when its *centre* is actually 50' below the horizon (alt $= -50'$).

Spherical Triangles and Early Devices for their Solution

Then $\sin(-50') = \sin 51 \sin 13 + \cos 5' \cos 13 \cos(\text{HA})$

$$\cos(\text{HA}) = \frac{\sin(-50') - \sin 51 \sin 13}{\cos 51 \cos 13} \qquad (50' = 0°.8333)$$

$(\text{HA}) = 107.99.$

The hour angle is therefore increased on account of refraction by $107.99 - 106.57 = 1°.42 = 5.68$ minutes.

The length of the day is therefore increased by 2×5.67 minutes $= 11.343$ minutes or *11 minutes 21 seconds*.

Sunrise will be at $(\text{HA}) = 107.99$ or $7^\text{h}.2$ before true sun noon 7^h 12^m or at 04^h 48^m approx. Sunset at 19^h 12^m.

These results ignore the slight change in the Sun's declination which it undergoes between sunrise and sunset, and they give the total sun time which has to be corrected to bring it to GMT. True sun time = local mean time + equation of time (from *Whitaker's Almanack*). This will be dealt with more fully in Chapter 9.

Refraction is about 34' when observing celestial objects on the horizon but rapidly decreases until at an altitude of 10° the effect of refraction is only about 5' and at 45° it is down to 1' (see Section 4.7(2)).

In Fig. 3.25 (1) represents the 'theoretical sun' with semi-diameter 16' regarded as a celestial body observed at zero altitude (zenith distance 90°) free from atmospheric refraction.

(2) represents approximately the positions of the 'true sun' (b) and of the 'apparent sun' (a) when observed through the Earth's atmosphere which increases the apparent altitude of the Sun by about 34'. When we see the Sun wholly above the horizon it is in reality wholly below!

(3) (b) represents the Sun as it appears at sunrise (or sunset) with its centre 16' below the horizon, but the Sun's true position is as shown in (b) with its centre $16' + 34' = 50'$ below the horizon.

The diagram—Fig. 3.25—will help to explain how it is just possible for an eclipse of the Moon to be observed when both the Sun and the Moon are above the horizon. In this peculiar set of circumstances the Sun, Earth and Moon are in a straight line, but the refraction of the atmosphere makes them appear to be making an angle of about 179°. The Sun and Moon have both actually set, by their true astronomical positions, but refraction of the atmosphere renders them both visible from the Earth.

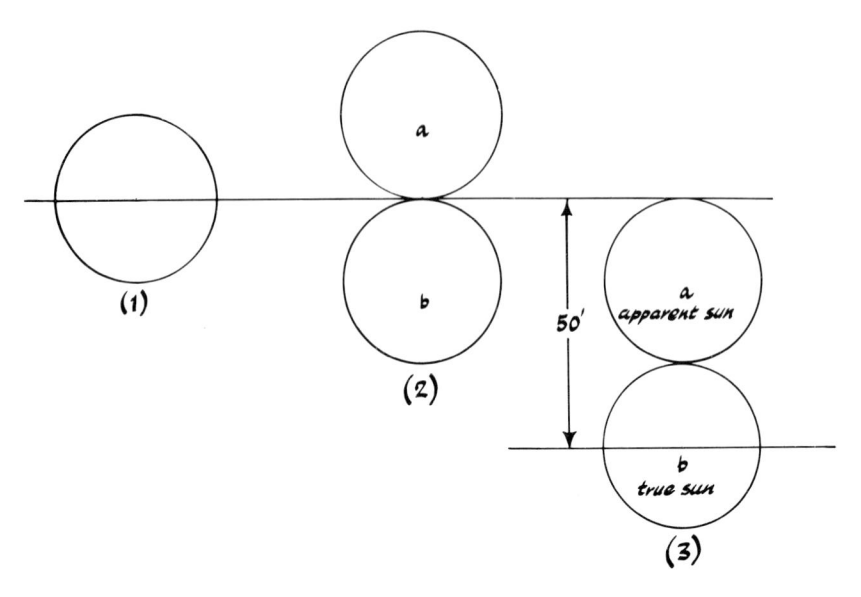

Fig. 3.25

3.25 How Much Sunshine can be Expected?

It is sometimes a matter of interest for travellers and most holiday-makers to know how much sunshine a particular place can have at a particular time of year. Yachtsmen find it useful to know the exact time of sunrise and sunset, as will be explained in Chapter 4.

This question entails finding the hour angle at which the Sun rises and sets at the time of year.

The relation to use here is [3.7(1)] and we can avoid a rigorous treatment of refraction of the atmosphere which in effect increases the length of a day's sunshine by about 11 minutes.

Fig. 3.26. Graphs showing how times of rising and setting of celestial bodies vary with the latitude of the observer and the declination of the body.

The curves are derived from relation [3.7(1)];

$$\sin(\text{alt}) = \sin\phi \sin\delta + \cos\phi \cos\delta \cos(\text{HA})$$

putting alt = 0 for rising and setting. Then

$$\cos(\text{HA}) = -\tan\phi \tan\delta.$$

The tables are calculated from this relation using the calculator.

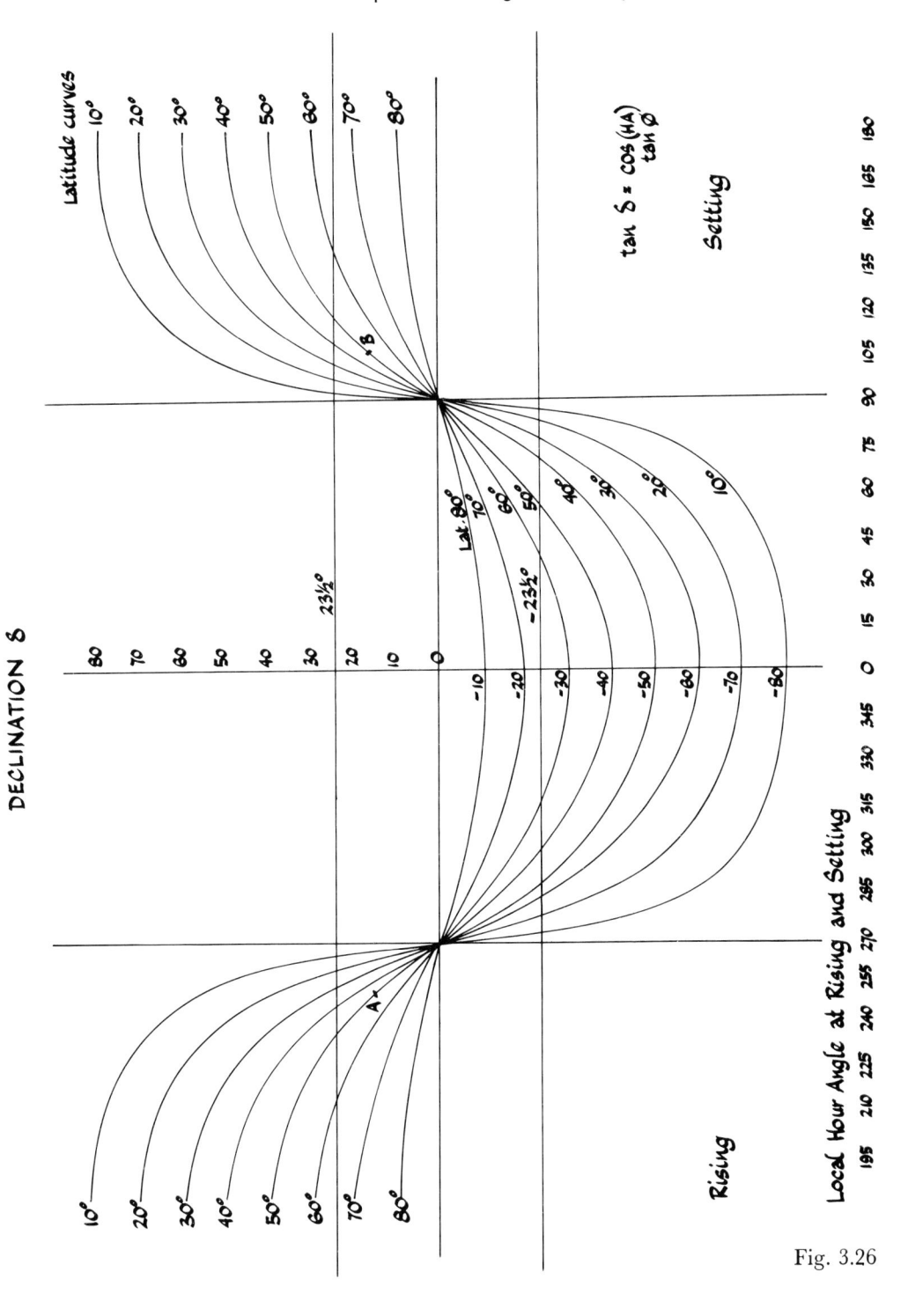

Spherical Triangles and Early Devices for their Solution

Fig. 3.26

$$\sin(\text{alt}) = \sin\phi \sin\delta + \cos\phi \cos\delta \cos(\text{HA})$$

When (alt) = 0 at sunrise or sunset,

$$\cos(\text{HA}) = -\tan\phi \tan\delta.$$

It will be found a rewarding project to draw up a table, with the help of the calculator, as shown in Table 3.1.

When using $\tan\delta = -\cos(\text{HA})/\tan\phi$ calculator users will discover that time can be saved by putting $-1/(\tan\phi)$ in the 'memory' for a run of HA's, i.e., for HA being given the values 15°, 30°, 45°, 60°, 75°, 90°. The table of figures is symmetrical about (HA) = 0. Values (HA) = 345°, 330°, 315°, etc., are the same as for (HA) = 15°, 30°, 45°, etc.

The curves showing the way the times of rising and setting vary with declinations and latitudes make a pleasing pattern (Fig. 3.26).

To find how much sunshine is possible on any day, mark two points on the appropriate latitude curve where the curve is cut by the δ ordinate. Measure the distance between the two points to find the time that the Sun is visible. For example, A, and B on latitude 50 at declination 15°N, total sunshine = 14.3 hours.

The curves also show under what conditions of latitude and declination a star is circumpolar, or under what conditions it can never rise or set. For example, a star of declination 45°S will not set if observed at latitude 50. The Sun at mid-summer does not set in latitudes greater than 66°.5 and does not rise in latitudes greater than 66°.5 in mid-winter.

Coordinates calculated from the rising and setting formula,

$$\cos(\text{HA}) = -\tan\phi \tan\delta.$$

To find how long the Sun or a particular star or planet will be above the horizon in say latitude 60° with a declination of 8°.5N.

(1) Look down the latitude 60° column till you come to the required declination, 8°.5N.

(2) Note the HA of the body's rising −105° (measured from 0°). This will also be the HA of setting.

Total time above horizon expressed in degrees = 2 × 105 = 210°. Expressed in hours = 210/15 = 14 hours.

Spherical Triangles and Early Devices for their Solution

Table 3.1 Table showing how the hour angle of the Sun, or a particular star or planet at the moment of its rising or setting varies with its declination. This relationship is shown for various latitudes.

HA	$\phi = 10$	$\phi = 20$	$\phi = 30$	$\phi = 40$	$\phi = 50$	$\phi = 60$	$\phi = 70$	$\phi = 80$
0	−80.00	−70.00	−60.00	−50.00	−40.00	−30.00	−20.00	−10.00
15	−79.65	−69.35	−59.13	−49.02	−39.00	−29.15	19.37	9.66
30	−78.49	−67.20	−56.31	−45.90	−36.00	−26.56	17.50	8.68
45	−76.00	−62.76	−50.77	−40.12	−30.70	−22.21	14.43	7.11
60	−70.57	−53.95	−40.89	−30.79	−22.76	−16.10	10.31	5.04
75	−55.73	−35.42	−24.15	−17.14	−12.25	−8.5	5.38	2.61
90	0	0	0	0	0	0	0	0
105	55.73	35.42	24.15	17.14	12.25	8.5	5.38	2.61
120	70.57	53.95	40.89	30.79	22.76	16.10	10.31	5.04
135	76.00	62.76	50.77	40.12	30.70	22.21	14.43	7.11
150	78.49	67.20	56.31	45.90	36.00	26.56	17.50	8.68
165	79.65	69.35	59.13	49.02	39.00	29.15	19.37	9.66
180	80.00	70.00	60.00	50.00	40.00	30.00	20.00	10.00

The graphs in Fig. 3.26 were drawn from the data in Table 3.1.

3.26 Sunrise, Sunset and Twilight

A note on sunrise and sunset and when best to observe stars with a sextant is included here and will be referred to in the next chapter. The times for sunrise and sunset are given in most almanacs for various latitudes and can be calculated as in Section 3.24. The published times of sunrise and sunset are calculated for the appearance of the Sun's upper limb and takes into account the equation of time as well as refraction.

Sunrise occurs when the Sun's upper limb is approximately 32' below the horizon, and therefore the centre is in fact $32' + 16' = 48'$ below the horizon, i.e., when the tip of the upper limb of the Sun is on the horizon the altitude of the Sun is minus 0°.8. Therefore we can use the relation,

$$\cos(\text{HA}) = \frac{\sin(\text{alt}) - \sin\phi \sin\delta}{\cos\phi \cos\delta}$$

to find the Sun's hour angle at rising or setting.

Subtract the hour angle from 12 and we get the 'local sun time of rising'. Now apply the equation of time and we get the time required.

An example for 18th November 1976, latitude 52°, will suffice to show the use of the calculator in finding sunrise and sunset times, LMT,

$$\cos(\text{HA}) = \frac{\sin(-0°.8) - \sin 52° \sin -19°.2}{\cos 52° \cos 19°.2}.$$

The calculator gives (HA) = 65°.057, which is 4 hours 20.2 minutes. Hence sunrise is at $12 - 4^h 20.2^m = 7^h 39.8^m$ and sunset at $12 + 4^h 20.2^m$. The equation of time on this date is $14^m 52^s$ (positive). Hence time of sunrise is $7^h 25^m$ LMT and sunset $16^h 05^m$ to the nearest minute.

Most countries have a legal definition of 'lighting up time' for traffic. This is generally about half an hour after sunset or half an hour before sunrise, but the actual moment depends on the latitude and longitude of the place.

Twilight also has to be defined. This is done by recognising three grades of twilight:

(1) *Civil Twilight*. This begins or ends when the Sun's centre is 6° below the horizon, and marks the time when operations requiring daylight should cease. It lasts from 30 to 60 minutes in latitudes around 50°.

(2) *Nautical Twilight*. This begins or ends when the Sun's centre is 12° below the horizon, and marks the time when for practical purposes it is dark, and generally the horizon as well as stars can be seen. This is a good time for taking sights of stars or planets.

(3) *Astronomical Twilight*. This begins or ends when the Sun's centre is 18° below the horizon, and marks perfect darkness, e.g., the horizon is not visible.

Reference to the *Nautical Almanac* will show that Edinburgh enjoys nautical twilight in mid-summer all night long. In London (lat 51°) in mid-summer, the Sun dips below the horizon by $90° - 51° - 23\frac{1}{2}° = 15\frac{1}{2}°$, so that London in mid-summer is in astronomical twilight all night but fails to get into the nautical twilight grade all night. The Shetland Islands (lat 61°N) has civil twilight all night in the summer.

4
Astro-Navigation with a Calculator

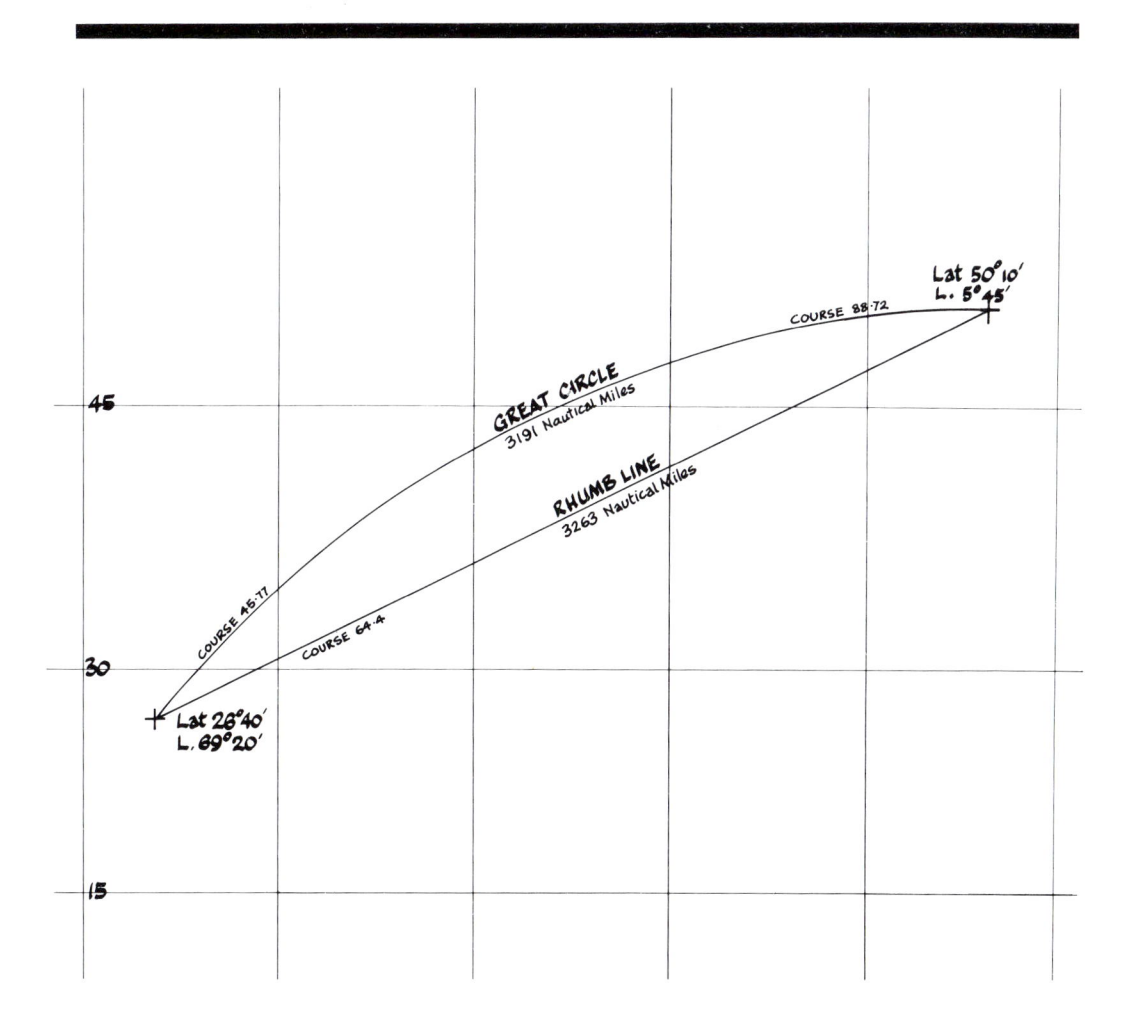

4.1   The development of positional astronomy is closely linked with the development of navigation. The *Nautical Almanac and Astronomical Ephemeris* considered as 'the seaman's bible' was conceived over two hundred years ago (when sextants and good time-keepers came into general use) with the purpose of determining longitude using lunar tables; but in the latter part of the 19th Century the *Nautical Almanac* published general essential data both for astronomical navigation and for astronomers.

It may be helpful, before using the calculator to solve the astro-navigation spherical triangle PZX (Figs. 2.2 and 3.13, formed by the pole P, observer's zenith Z and the star X), to bring the problem down to earth and consider a comparable spherical triangle on the Earth's spherical surface. Given three points on the Earth's surface, A the Pole, the latitudes and longitudes of, say, an ocean yacht C and its remote destination B, we can find by calculation the distances and directions between these three points, as shown in the following example (Fig. 4.1).

GREAT CIRCLE SAILING OR FLYING

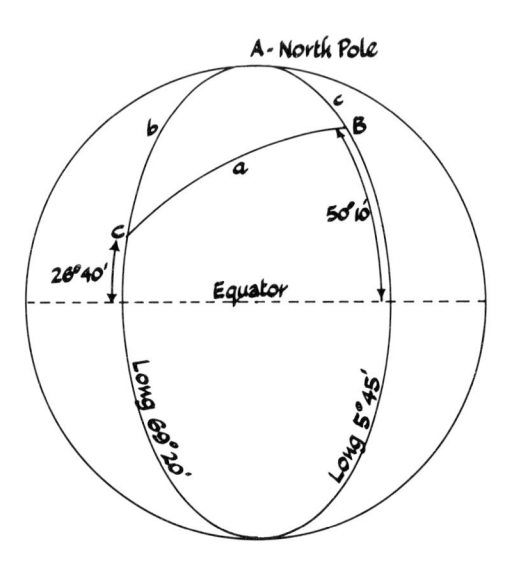

Fig. 4.1. Great circle sailing or flying. C represents an aircraft's or yacht's position in the Atlantic, lat 26°40'N and long 69°20'W. B represents a point off the coast of Cornwall, lat 50°10'N and long 5°45'W. a = great circle distance from C to B. a = 53°.1842 or 3191'
= 3191 nautical miles.

4.2 Great Circle Sailing

A yachtsman in the Atlantic not sure of his position, seeks help from a friendly passing ship and gets his position as lat 26°40′N, long 69°20′W. He wishes to calculate how far he is from a point on the Cornish coast, lat 50°10′, long 5°45′. This distance he can calculate in less than a minute.

Figure 4.1 shows the positions of the North Pole, A of the Earth, C the yacht's position, lat 26°40′N, long 69°20′W. B is the coast of Cornwall, lat 50°10′, long 5°45′.

In the triangle ABC we know the following:

Angle at A, the Pole, is the difference in longitude between the yacht at C and the Cornish coast at B

$$= 69°20'W - 5°45'$$
$$= 63°35'.$$

The side b is (90°—latitude of yacht).

The side c is (90°—latitude of the Cornish coast).

The side a is the side to be calculated in degrees and minutes of arc, and is part of a 'great circle'.

The most convenient unit for expressing distances between two points on the surface of the Earth, is that defined by the great circle distance between two points subtending an angle of one minute of arc at the centre of the Earth. This unit in common use in navigation is known as the 'nautical mile', and is taken to be 6080 feet or 1.853 km. There are small variations in this value as the earth is not a perfect sphere.

The great circle distance in nautical miles is simply degrees and decimals of a degree $\times 60$.

The formula for solving the spherical triangle is from Fig. 3.11:

$$\cos a = \cos b \cos c + \sin b \sin c \cos A.$$

Now in order to use latitudes and longitudes we use the transformation above which in effect changes cosines into sines and vice versa, through their complementary angles, as was shown in Section 3.5.

We can therefore write the equation of Fig. 3.11 as:

$$\cos a = \sin(\text{Lat of yacht}) \sin(\text{Lat of Cornish coast})$$
$$+ (\cos(\text{Lat of yacht}) \cos(\text{Lat of Cornish coast})$$
$$\times \cos(\text{difference in Lat between yacht and Cornish coast})).$$

Putting in the values for latitudes and longitudes we have,
$$\cos a = \sin 26°.666 \sin 50°.166 + \cos 26°.666 \cos 50°.166 \cos 63°.583.$$

The pocket calculator gives the answer in less than a minute of time:
$$a = 53°.184$$

which is multiplied by 60 to give minutes of arc, or nautical miles.

The great circle distance—the shortest distance in fact is 3191 nautical miles.

The rhumb line distance is 3263 nautical miles, see Section 4.14.

As a matter of interest, if we had taken the spherical triangle ABC in Fig. 4.1 as a plane triangle and applied the cosine formula for plane trigonometry, $a^2 - b^2 + c^2 - 2bc \cos A$, we would have:
$$a^2 = (3814)^2 + (2390)^2 - (2 \times 3814 \times 2390 \times \cos 63.583).$$

The calculator gives in about 1 minute of working time, $a = 3484.8$ nautical miles, which is 284 miles greater than the spherical triangle, or great circle distance.

The sides joining ABC of a spherical triangle are the shortest distances between A, B and C. They are parts of *great circles*.

4.3 For those who are not familiar with the scientific calculator the following detailed programme is given showing the display at each stage.

Most scientific calculators follow an algebraic logic system which is carried out below.

In practice no figures need to be written down at all, apart from the initial formula—which is really only a programme to be followed in numbered sequence, 1, 2, 3 to 20, etc.

```
              (2)    (1)    (3)(5)   (4)    (6) (7) (9)   (8)    (10)(12)   (11)
cos a  =   sin 26.666 × sin 50.166 + (      cos 26.666   ×  cos    50.166

                                                              (13)  (15)  (14)   (16)
                                                                ×         cos 63.583  )

           (17) (18)   (19) (20)
         = arc cos  ×   60
```

Result 3190.8 nautical miles.

Why do we put in the brackets? It is simple to tell the calculator precisely what we want it to do, e.g., $4 + 3 \times 2$ could be $4 + (3 \times 2) = 10$ or it

Astro-Navigation with a Calculator

might be $(4+3) \times 2 = 14$. (The calculator can't think or read your thoughts—it only obeys instructions!)

Operation	Enter figure	Depress function key	Read on display	Notes
Start	—	c (clear)	0	
1*	26.666		26.666	*The numbers correspond to the numbers in brackets over the programme above.
2		sin	0.448 788	
3		×	0.448 788	
4	50.166		50.166	
5		sin	0.767 903	
6		+	0.344 626 49	This is the product of sin 26.666 and sin 50.166 and is stored.
7		(	0	The bracket isolates temporarily the cosine products *and* puts the 0.344 626 49 in store, to be used at 17.
8	26.666		26.666	
9		cos	0.893 637 8	
10		×		
11	50.166		50.166	
12		cos	0.640 654 5	
13		×	0.572 505 4	
14	63.583		63.583	
15		cos	0.444 900 9	
16		)	0.254 676 2	This closes the bracket opened at 7.
17		=	0.599 302 7	This is the sum of 6 and 16.
18		arc cos	53.180 0	This is the angle in degrees.
19		×	53.180 0	This produces *a* in minutes of arc.
20	60		3190.8	This is distance in minutes of arc or nautical miles.

4.4 After the example of Section 4.2 using the spherical triangle on the surface of the Earth, we now return to the celestial sphere. Figure 4.2 shows the navigational spherical triangle *ABC* formed by the North Pole, *A*, our own position *C* marking the zenith (all observers are apparently on top of the world!), and the position, *B*, of the star we are observing. This triangle is often referred to as the *PZX* triangle.

A is the local hour angle.

a is the zenith distance of the star, which is the co-altitude of the star, or (90−(alt)).

b is the co-latitude of the observer at *C*.

c is the co-declination of the star, or (90−δ).

81

The basic relation for spherical triangles thus takes on the form that is most useful in positional astronomy and particularly in astro-navigation:

$$\sin(\text{alt}) = \sin \delta \sin \phi + \cos \delta \cos \phi \cos(\text{LHA}).$$

To take a simple example of a star δ 12°, latitude of observer 51°, local hour angle of star 73°, the altitude is given by:

$$\sin(\text{alt}) = \sin 12° \sin 51° + \cos 12° \cos 51° \cos 73°$$

Altitude is 19°.971 5.

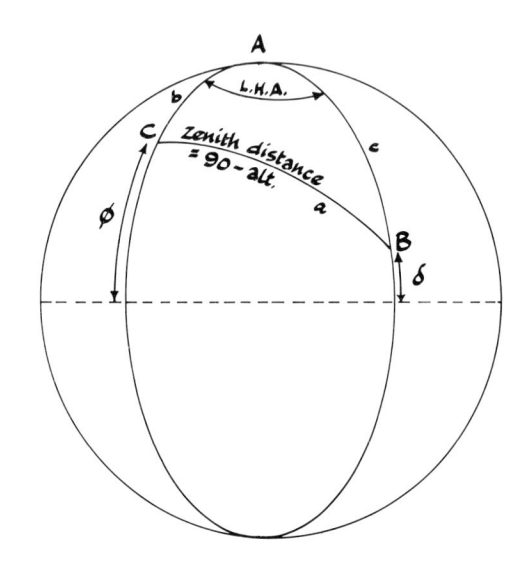

Fig. 4.2. Astro-navigation. C represents the position of the observer (estimated), lat 51°14′N and long 2°50′W. Coordinates of star Arcturus, declination $\delta = 19°18.7′$, local hour angle $= 63°30.8′$. Calculated altitude $= 31°25.8′$. For detailed calculations see Section 4.6.

4.5   To Find Your Position by Observing the Altitudes of Two Stars, Using a Calculator, a Good Watch and *Whitaker's Almanack*, or a *Nautical Almanac*, in 11 Steps

(It is generally more convenient to use the *Nautical Almanac*. See (4) below.)

(1) Find the altitude H of a selected star by sextant, e.g., Arcturus, and note the exact GMT.

(2) Apply the necessary corrections to the sextant reading for atmospheric refraction, index error, and height above horizon if using a sea

horizon, see Section 4.7. Corrected $H = H_0$. This H_0 is the true altitude of the star observed from your actual position.

(3) Make an *estimate* of where you think you are. Latitude ϕ, longitude L.

(4) From an almanac or astronomical yearbook, for the date and time, and using your estimated longitude, find the star's LHA using sidereal time at 0^h, the RA* and GMT of observation, as in Section 2.9.

$$(\text{LHA})^* = \text{sidereal time at } 0^h + \text{GMT} \times (1.002\,738) - (\text{RA})^* {}^{+E}_{-W} \text{Long}$$

Express the LHA in degrees and decimals of a degree.

(When using the *Nautical Almanac* the LHA of the star is equal to (GHA) ♈ (at time of observation) $+ (\text{SHA})^* {}^{+E}_{-W}$ Long.)

(5) Look up the declination of the star δ, from the almanac.

(6) Calculate what the altitude H_c would be if the star were actually observed from the estimated position (ϕ, L), i.e., use the spherical triangle relation [3.7(1)].

$$\sin H_c = \sin \phi \sin \delta + \cos \phi \cos \delta \cos(\text{LHA}).$$

Write down H_c from the display panel of the calculator.

(7) $H_0 - H_c$ tells you how much nearer the star (in the direction of the star) you really are than you estimated. (1' arc = 1 nautical mile.) This direction is the azimuth of the star. Denote azimuth by Az.

(8) This azimuth of the star from the estimated position is important and is found in a few seconds from the relation,

$$\sin(\text{Az}) = \frac{\sin(\text{LHA}) \cos \delta}{\cos H_c}$$

see equation [3.7(7)].

(9) Draw a line AB on your chart or map through your estimated position towards the star, i.e. with Azimuth as found in (8).

(10) Draw a line perpendicular to AB but at a distance $H_0 - H_c$ (in nautical miles) from your estimated position. This perpendicular should be drawn nearer to the star from the assumed position if $H_0 > H_c$, but *away* from the star if $H_0 < H_c$. This line, CD, is *the position line from the star*. You are somewhere on it.

(11) Repeat these operations (except for (3) which is generally the same) for another star and so obtain a second position line RS. You are also on this line. Therefore, your true position is at D, the point of intersection of CD and RS (see diagram of actual example Fig. 4.3).

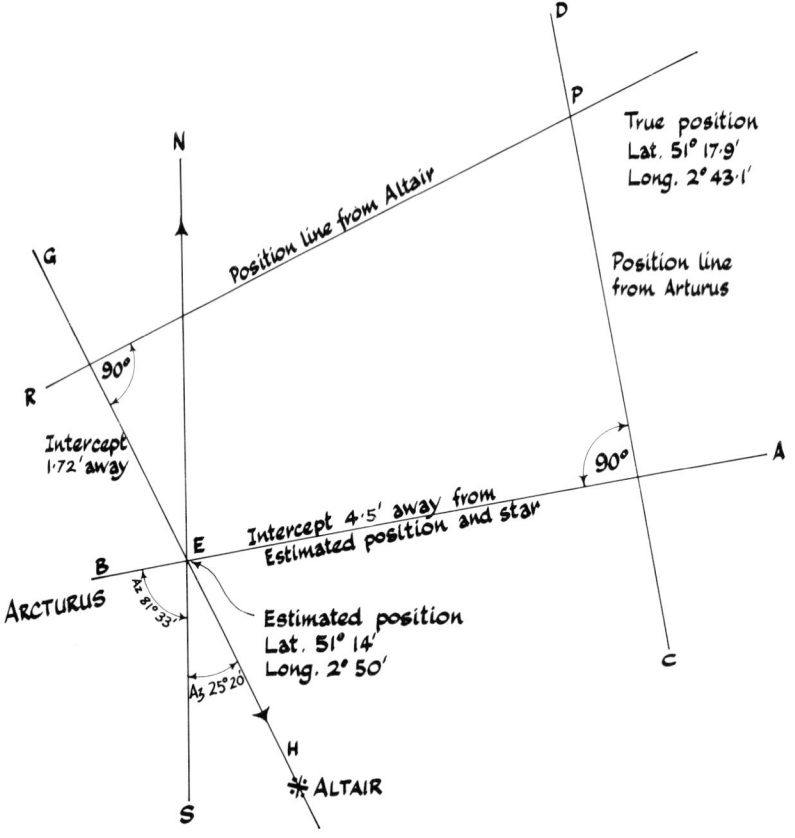

Positional Astronomy and Astro-Navigation Made Easy

Fig. 4.3.

In the diagram the line GH is the line through the estimated position drawn in the calculated direction of the second star and RS is the line perpendicular to it. It is drawn at a distance of $H_0 - H_c$ away from the star, following the same procedure as in (10).

Apply the steps (1) to (10) to an actual observation, for example on the star Arcturus at 29th August 1975, $20^h\ 10^m\ 15^s$.

(1) Altitude of Arcturus by sextant (corrected reading),
$$H_0 = 31°21' = 31°.35.$$

(2) GMT of observation $20^h\ 10^m\ 15^s$ or 20.170 833 hours.

(3) Estimated position, latitude $\phi = 51°14'$ or $51°.233\,333$, longitude $L = 2°50'W$ or 0.188 88 hours.

Astro-Navigation with a Calculator

(4) From *Whitaker's Almanack*, sidereal time at 0^h on date = $22^h\ 26^m\ 19^s$ or 22.438 6 hours.

Right Ascension of Arcturus (RA) = $14^h\ 14^m\ 30^s$ or 14.241 67.

Local(HA)* = sidereal time at 0^h+GMT (1.002 738) − (RA) −longW

$= 22.438\ 61 + 20.170\ 833\ (1.002\ 738) - 14.241\ 67 -$
$\quad -0.1888\ 8$
$= 4.234\ 201$ hrs or $63°.513\ 01$.

(5) Declination of Arcturus+19°18.7' or 19°.311 67. Apply the relation [3.7(1)].

(6) $\sin H_c = \sin 51.233\ 3\ \sin 19.311\ 67 + \cos 51.233\ 3\ \cos 19.311\ 67\ \cos 63.513$

whence $H_c = 31°.426\ 17$.

(Time on the calculator, 64 seconds.)

(7) $H_0 - H_c = 31°.35 - 31°.426\ 17 = 0.075°$ ($H_0 < H_c$ ∴ away)
$= -4'.5$.

This means that our position line is 4.5 nautical miles *away* from the star from our estimated position.

(8) The *direction* of the star (Az) is given by,

$$\sin(\text{Az}) = \frac{\sin 63.513\ 01\ \cos 19.311\ 67}{\cos 31.35}$$

(Az) = 81°31' from the *S* point to the West.

Therefore the true bearing of Arcturus is 261°.31, (81°31'+180).

(9) Draw the azimuth line at 81°31' from the meridian, S81°31'W.

(10) *Draw the position line at right angles to the azimuth line and 4'.5 away from star* (see Fig. 4.3).

Apply the steps 1–10 to an observation of Altair.

(1) Altitude of Altair by sextant. Corrected reading, H_0 45°5.3' = 45°.083.

(2) Time of observation 29th August 1975, $20^h\ 20^m\ 5^s$ = 20.334 7 hrs.

(3) Estimated position: as for the Arcturus observation.

latitude, 51°14' = 51°.233 3
longitude, 2°50W = 0.188 8 hrs.

(4) From *Whitaker's Almanack*, sidereal time at 0^h on date, as before for Arcturus = 22.438 6 hrs.

(RA) Altair = 19^h 49.6^m or   19.826 66 hrs.

$(LHA)^*$ = sidereal time at 0^h+GMT(1.002 738)−(RA)$_W^E$Long
= 22.438 6+20.334 7(1.002 738)−19.826 66−0.188 8
= 22.813 516 hrs
= 342°.203

(5) Declination of Altair = N.8°48.5′ = +8°.808 3

(6) $\sin H_c$ = sin 51.233 3 sin 8.808 3 + cos 51.233 3 cos 8.808 3 cos 342°.203. By calculator, H_c = 45°.117 08 = 45°7′.02 (time for the calculation, 60 seconds).

(7) $H_0 - H_c$ = 45°5′.3 − 45°7′.02 = −1.72   ∴ 1.72 miles *away*.

(8) Azimuth from $\sin(Az) = \dfrac{\sin 342.203 \cos 8.803\,3}{\cos 45.083}$ and $\begin{array}{l}Az = -25°.326 \\ = -25°20′\end{array}$

This measured from the S point of the meridian gives true azimuth, 180 − 25°20′ = 154°40′.

(9) Draw the azimuth line from (8) *GH*.

(10) Draw position line perpendicular to azimuth line, 1.72 N. miles *away*. Position is at the point of intersection of *CD* and *RS*.
See Section 4.18 for a further example and a method of recording data.

4.7 Corrections to be Made to the Sextant Reading

When finding the altitude of a celestial body by means of a sextant there are many precautions to be taken, and corrections to be applied for maximum accuracy, and it must be emphasised that the use of electronic calculators does not reduce the need for careful planning and observing. The main corrections are as follows.

Dip. The calculator makes the calculation of dip very easy as it can deal with angles so small that they are not amenable directly to four figure log and trig tables.

A navigator at O taking the altitude of a star from a height of, say, h above sea level views the horizon at an angle slightly lower than the true horizontal (Fig. 4.4). The horizon appears to be dipped below the horizontal by a few minutes of arc which can readily be calculated and allowed for.

Astro-Navigation with a Calculator

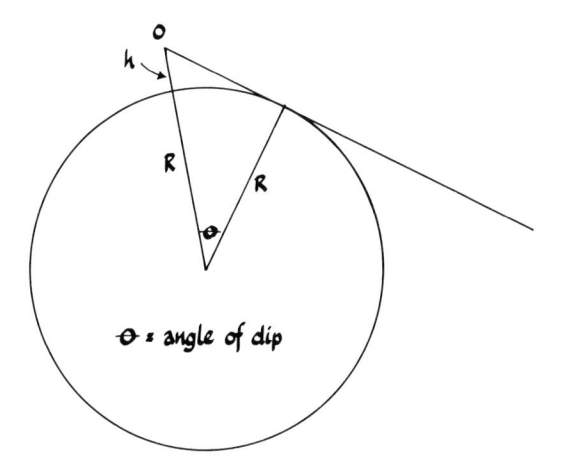

Fig. 4.4

R is the Earth's radius and is 6378 km.

From the diagram the angle between the true horizontal and the direction of the horizon is the angle θ, and

$$\cos \theta = \frac{R}{R+h}. \qquad [4.7(1)]$$

To find θ the angle of dip to be *subtracted* from the sextant reading, for a height above sea level of 3 m or 0.003 km, we have neglecting the small effect of atmospheric refraction from the horizon to the observer.

$$\cos \theta = \frac{6378}{6378.003}. \qquad \text{(This cannot be worked out conveniently by log tables.)}$$

The calculator gives at once,

$$\theta = 0°.055\ 573\ 86 \quad \text{or} \quad 3'.334$$

The angle of dip = 3'.334 of arc (which agrees with the approximate formula, dip = $1.06 \sqrt{\text{ht in feet}}$ or $1.92 \sqrt{\text{ht in metres}}$.

As an exercise in the use of the 'dip' relation the following question which was set for fun in an astronomical journal might be of interest.

"How high must you go up in a balloon in order to receive 13 hours of sunshine at lat 14 in October (Sun's dec $+8°.75$)?"

The HA at rising and setting must be $\frac{13}{2}$ = 6.5 hours, or 97°.5, so the *altitude* under these conditions must be given by

$$\sin(\text{alt}) = \sin 14° \sin 8°.75 + \cos 14° \cos 8°.75 \cos 97°.5.$$

The altitude calculated is $-5°.07$. The minus sign indicates that for

HA 6.5^h ($97°.5$), $\delta = 8.75$, $\phi = 14$ the Sun will be $5°.07$ *below* the horizon.

Therefore $\qquad \cos -5°.07 = \dfrac{6378}{6378+h} \quad$ so $\quad h = 6378\left(\dfrac{1}{\cos 5°.07} - 1\right)$

$h = 25$ km (this ignores the effect of refraction of the atmosphere).

A convenient way to evaluate this expression for many types of calculator is:
(1) find cos 5°.07, note cos 5°.07 is the same as cos $-5°.07$,
(2) then the reciprocal,
(3) subtract 1,
(4) multiply by 6378.

With practice you will discover many short cuts and time saving operations depending on the type of calculator used.

Atmospheric Refraction. This effect is not amenable to any simple theoretical formula as it depends on meteorological conditions, temperature and pressure, but for practical purposes the effect is about 5' at low altitudes, around 10°. At altitudes below 6° sights are regarded as unreliable. A correction for refraction should however be applied to all sights, but at altitudes greater than 50° the correction becomes less than 1'. All corrections are subtracted from the sextant reading. A graph is useful for quick reference showing roughly how atmospheric refraction varies with altitude of the body.

Correction for Index Error of Sextant. This merely compensates for the fact that when viewing the horizon, so that it appears in a continuous unbroken horizontal line, the scale reading on the sextant may not be exactly at zero. The amount of this 'off zero' reading is the index error. If it is minus then all its readings will be low by this amount, and so the index error must be added. If it is positive then the index error must be subtracted.

The Semi-Diameter of the Sun or Moon. Positional astronomy concerning the Sun, Moon and planets defines positions of the centres of these bodies, so, when the Sun's or Moon's lower limb is 'brought to the horizon' it is clear

---

Fig. 4.5. Graph showing relation between refraction of atmosphere and altitude. Correction to be substracted. When altitude is 90° refraction is 0°. When altitude is 0 the refraction is large and = 34' (approximately the diameter of the Sun or Moon). The Sun appears wholly above the horizon, when it is really wholly below. Avoid low altitude. Turbulence and temperature gradients can affect the refraction figures by small amounts.

Astro-Navigation with a Calculator

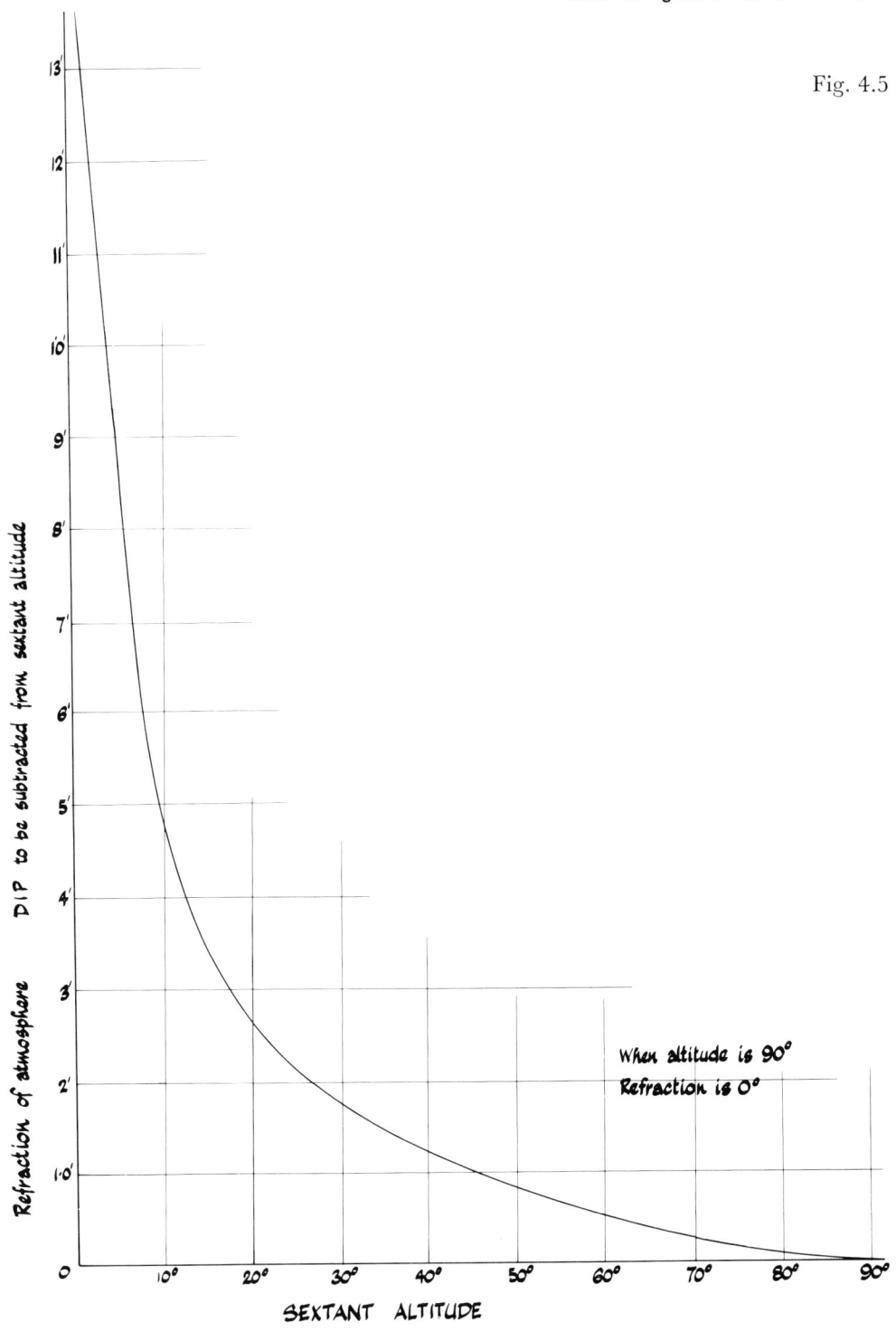

Fig. 4.5

that the altitude angle measured is not large enough by the semi-diameter of the body, so the semi-diameter must be added to the sextant reading. Semi-diameters of Sun and Moon vary slightly from day to day depending on their distances from the Earth, but the semi-diameters are given every few days in the *Nautical Almanac*, and each month in *Whitaker's Almanack*.

Timing of Observations. Accurate timing of observations is necessary and should present no great problem these days when time signals by radio are readily available. A stop watch is an essential adjunct to the timing equipment as it enables an observer to record an exact time of observation. An error of 4 seconds can create an error of 1 mile in plotting a position (see Section 10.3).

Note on Moon Sights—Parallax. Finding one's position from an observation of the Moon is a little more exacting than from observations of the Sun, as the Moon's declination and Right Ascension change much more rapidly than for the Sun, but with careful timing of the observations and strict regard to the increments and corrections tabled in the *Nautical Almanac* a Moon sight is a useful and pleasing observation, and the procedure is the same as for the Sun, with one important exception.

The Moon is much nearer to the Earth than the Sun is, so that the direction of the line joining the centre of the Moon to the centre of the Earth, i.e., a line exactly overhead from the observer at O, cannot be considered as parallel to a line joining the observer to the Moon from any other point on the Earth a few thousand miles away at B. This effect is known as parallax. The further away from the overhead point (from which the altitude of the Moon is 90°) the greater the effect of parallax. The parallax is a maximum when the altitude of the Moon is 0°, or when the Moon is on the observer's horizon. This maximum effect angle X is called the 'horizontal parallax' (HP) and can easily be calculated using the figure opposite (Fig. 4.6). R is the radius of the Earth. D is the distance of the Moon from the centre of the Earth.

For an observer at A, the horizontal parallax (HP), angle x is given by

$$\sin x = \frac{R}{D} = \frac{6.38 \times 10^6}{3.844 \times 10^8}$$

$$= 0°.95 \quad \text{approx.}$$

This varies a little from day to day because D varies, on account of the Moon's slightly elliptical orbit around the Earth.

Astro-Navigation with a Calculator

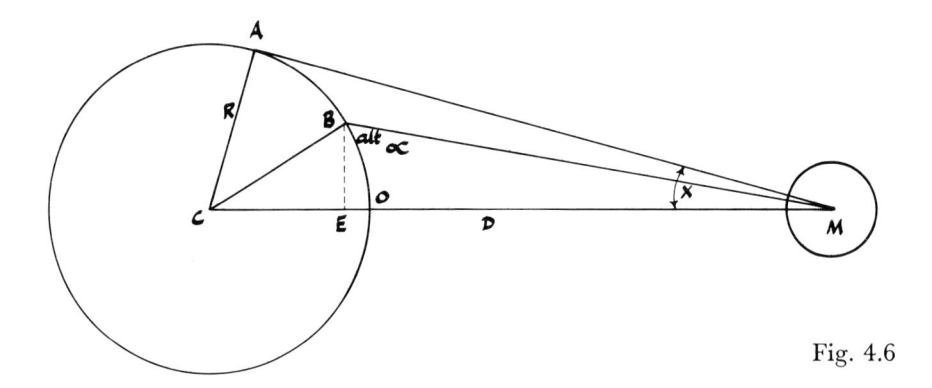

Fig. 4.6

For an observer at B who observes the Moon at an altitude of a, the approximate parallax is,

$$\frac{BE}{CM} = \frac{R \cos a}{D} = (HP)\cos a$$

and this angle must always be added to the observed altitude.

The horizontal parallax is given for each day in the *Nautical Almanac* and varies from about 61′ to 54′. So for an observation of the Moon, HP 60′, at an altitude of 30° the parallax correction would be 60 cos 30 = 52′, a very important correction.

There is a corresponding parallax correction to be made for an observation of the Sun (about 0′.13) but for our practical purposes this can be ignored.

Methods for recording and using data for drawing a position line, from a sextant altitude of stars, Sun and Moon, are suggested in Section 4.18. This is a departure from the more traditional lay-out for working sights, but the increasing use of the pocket scientific calculator calls for a fresh approach as paperwork is now minimal. Astro-navigators will doubtless devise their own systems, but Sections 4.18 and 4.20 contain the essential data.

4.8 Practice With a Sextant—on Land

By using a scientific calculator a great deal of useful practice in celestial navigation can be obtained in one's back garden, with the help of a cheap but reasonably accurate plastic sextant used in conjunction with an artificial horizon. This could consist of a small bowl of oil, treacle or glycerine, or even a good sheet of glass mirror carefully levelled by accurate spirit

Positional Astronomy and Astro-Navigation Made Easy

Fig. 4.7

Fig. 4.8. Star as seen through the telescope of the sextant reflected in the level glass plate.

Astro-Navigation with a Calculator

Fig. 4.9

levels and levelling screws (see Figs 4.7, 8 and 9). A useful artificial horizon for practice (used by the author) can be made from a sheet of good plate glass about 200 mm × 200 mm × 4 mm which has a ground or frosted surface on one side. This side is painted black with blackboard paint. The sheet is mounted on a wooden base fitted with three levelling screws consisting of ordinary nuts and bolts, which can be adjusted to ensure that the upper plane polished surface of the sheet is horizontal to within three or four minutes of arc. One turn of a levelling screw can move an image of the Sun by about a semi-diameter, and a $\frac{1}{4}$ of a turn can easily be detected on the spirit level. Reflections of the Sun or stars take place only at the upper surface of the glass sheet while light that enters the glass surface by refraction is lost through scattering and absorbtion at the blackened and ground surface. Although this type of artificial horizon is as accurate only as the spirit level, it can be as accurate as a bubble sextant and has a great advantage over a liquid artificial horizon in that it remains unaffected by the wind, and is easily portable.

It is reassuring after taking sights in the garden, and consulting the *Nautical Almanac* or *Whitaker's Almanack* for declinations and GHA's at GMT to find that your house (or holiday camp) is just where it should be on the map, within the limits of error of your sextant, and your skill as an observer.

Note, when using an artificial horizon, that the angle between the star or Sun, and its reflection is double the altitude. Corrections for the altitude must be made for refraction and sextant index error, but corrections for dip are not required, and as the sextant image of the Sun or Moon is brought into coincidence with the image reflected from the artificial horizon, the question of the semi-diameter of the Sun, or Moon, does not arise.

4.9 A Nine Figure Display on the Panel of a Calculator can be Misleading

The important principle to be observed in using a calculator with a 14 character capacity and a nine figure display such as the one illustrated in Fig. 1.1, is that the maximum precision to be expected is determined solely by the accuracy of the data input. This is commonsense. This accuracy, however, can be impaired by a poor use of a formula or failure to retain an adequate number of figures at some intermediate stages, but it can never be increased (see Section 1.6).

Although the calculator handles numbers of 9 figures rapidly, and just as easily as numbers of two or three digits, as the examples in Section 4.3 will show, it is misleading to use final results showing more figures than is justified by the data input.

In the slide-rule and log table days the number of significant figures was necessarily limited by these devices, but with the modern calculators extra figures require an insignificant amount of extra work. The calculator still affords scope for mathematical sense and commonsense. For further comments on computation and calculator techniques reference is invited to the section on 'Computation and Interpolation' in the Explanatory Supplement to the *Astronomical Ephemeris* and *Nautical Almanac*.

One source of interest and fascination is to vary the input of one of the functions used, by a very small amount, say 1°, and find the effect this has on the final result. This can be a useful as well as an interesting and practical exercise as it enables one to determine easily and rapidly the effects of small errors of observations (see Chapter 10).

4.10 The Automatic Digital Sextant

This is really computer aided astro-navigation. A standard sextant can be modified with connections to a 'data store unit' slung from the navigator's shoulder. When taking a sight in the normal way, the navigator presses a button on the sextant which automatically records the observed altitude, and the time. The two essential requirements for a position line. This is rather like recording the winner and his time in a photo-finish race! This is a great convenience as it relieves the observer of timing and reading verniers.

The final great advantage is that the data recorded can be transferred to the ship's computer, by connecting the two together. The computer then calculates, and prints out on a display terminal, the ship's position, which is also automatically displayed on the ship's chart at the same time.

This may be the most sophisticated equipment yet devised for computerised astro-navigation, but for navigators in small ships or on yachts the pocket calculator will give great satisfaction to the user who will gain through its use understanding and confidence in the basic principles of spherical triangles. No matter how automatic or sophisticated the calculating and plotting devices may be, the greatest attainable accuracy is limited finally by the accuracy and skill with which the sextant can be operated by human hands and eyes in taking a sight.

4.11 Polaris and the Calculator

One of the simplest ways of finding your latitude is to measure the altitude of the Pole Star (Polaris) which is not exactly at the celestial pole but 51' from it. (It has a (1977) declination of 89°9'.) The altitude of the celestial pole (if it could be observed) is the latitude of the observer's position. As the Pole Star revolves round the celestial pole we are liable to be in error over our latitude by as much as 51' too much, or 51' too little, depending on whether Polaris is just above the pole or just below it (Fig. 4.10).

Navigators have rather elaborate tables which give the necessary corrections to be made depending on the time of day and year, but for most practical purposes the correction can be made by the nocturnal (Fig. 2.6), and a calculator, without recourse to any tables at all.

Hold the nocturnal up to the pole, but turn the moveable edge, or knitting needle, so that it is in line, not with the pointers of the Plough, but with the end star of the handle of the Plough (Alkaid). This is the tip of the tail of the Great Bear. Estimate the hour angle, and use a transparent protractor to assist in this estimate. A line joining Alkaid and Polaris passes through the celestial north pole to within a few minutes of arc, as shown in Fig. 4.10, so the angle Polaris makes with the meridian at the celestial North Pole (the hour angle) is SPM or θ. This is the same as APM, the angle that Alkaid makes with the meridian. In making an observation of the altitude of Polaris, H_p, the latitude will be, as explained below,

$$H_p - 51' \cos \theta \qquad [4.11]$$

Positional Astronomy and Astro-Navigation Made Easy

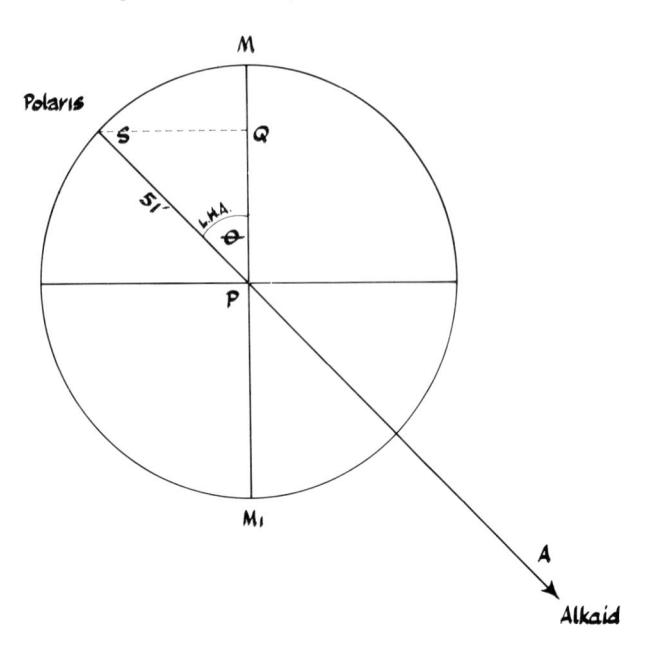

Fig. 4.10. S is the Pole Star, Polaris. P is the Celestial Pole, $SP = 51'$.

where θ is the local hour angle (see Fig. 4.10). When $\theta = 90°$ or $270°$ no correction need be made.

The results can be checked using tables given in the *Nautical Almanac*. Your estimate of the hour angle of Polaris by the nocturnal can be checked by finding the local sidereal time and subtracting the RA of Polaris ($2^h\ 8^m$) or $32°$. This is the same as adding the SHA of Polaris.

$$\text{LHA of Polaris} = (\text{LST}) - (\text{RA}) \text{ of Polaris}.$$

The result obtained using [4.11(1)] is approximate only as it assumes that the triangle SQP is a plane triangle, whereas it is a small spherical triangle, but the difference is insignificant unless extreme accuracy is required.

The altitude of S can be considered as the altitude of the celestial pole $P + PQ$. The altitude of P is the latitude of the observer, ϕ, and with $\theta = \text{LHA}$.

$$PQ = SP \cos \theta = SP \cos(\text{LHA}),$$

Latitude, ϕ = altitude of Pole Star $- 51' \cos(\text{LHA})$.

4.12 A convenient way of finding the hour angle of Polaris *accurately* is to:
(1) Look up the GHA of ♈ in the *Nautical Almanac* for the GMT, add

Astro-Navigation with a Calculator

Longitude W° or subtract Longitude E° to get the LHA ♈. (2) Add to this LHA ♈ the SHA Polaris. This gives the LHA*. Then the correction to be algebraically subtracted is 51′ cos(LHA) (the calculator will normally take care of the sign no matter what the size of the LHA may be).

4.13 Traverse and Departure

The hand held calculator can perform numerous simple but important calculations in navigation and makes it unnecessary to extract figures from volumes of tables such as *Traverse and Departure Tables*.

These tables have been compiled on the reasonable assumption that the distances involved are small (i.e., less than 600′ or about 10° on a great circle) and that positions and distances can properly be represented on a plane. Hence the expression 'plane' sailing which because of its simplicity is sometimes wrongly spelt 'plain' sailing.

In Fig. 4.11 ABC is one such plane triangle. A is a ship's present position (lat ϕ long L), and she sails to B a distance of S *nautical miles*. AC is a line of longitude, and AC is due north. The ship's course is $CAB = \theta$ ($BCA = 90°$).

The distance made good by the ship due North is $AC = S \cos \theta$. This is called D (lat). The new latitude is ϕ' and $S \cos \theta = \phi' - \phi$.

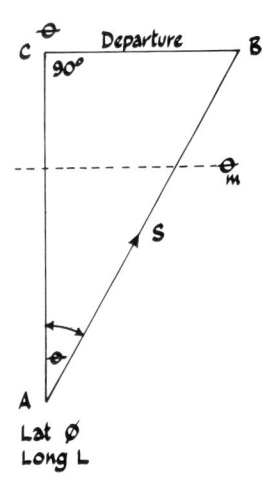

Fig. 4.11

The *distance* made good in an E–W direction, CB is $S \sin \theta$. This is called its 'departure' and is in nautical miles. To turn this distance into the difference in longitude between C and B we recall that 1° of longitude varies with the latitude being considered.

1 nautical mile measured E–W at lat ϕ will occupy $1/(\cos \phi)$ minutes of arc.

Therefore the difference in longitude between A and B is $\dfrac{S \sin \theta}{\cos \phi_m}$.

It will be appreciated that ϕ_m is between ϕ and ϕ', and a *mean* value $\phi_m = (\phi+\phi')/2$ is used in calculations and tables.

4.14 Example (Fig. 4.12)

What is the true course and distance to sail from A, lat 49°57′N, long 6°00′W, to B, lat 43°04′N, long 9°38′W? The mean lat $\phi_m = 46°$ approx (Fig. 4.2).

Let S be the distance in nautical miles and θ be the course from B to A. The difference in latitude is $49°57' - 43°04' = 6°53'$,

$$d(\text{lat}) = \tfrac{S}{60} \cos \theta = 6°53' = 6°.883 \qquad [4.13(1)]$$

$$BC = \tfrac{S}{60} \sin \theta = 3°38' \times \cos \phi_m = 3°.633 \times \cos 46°. \qquad [4.13(2)]$$

Divide [4.13(2)] by [4.13(1)]

$$\tan \theta = \frac{3.633 \times \cos 46}{6.883}$$

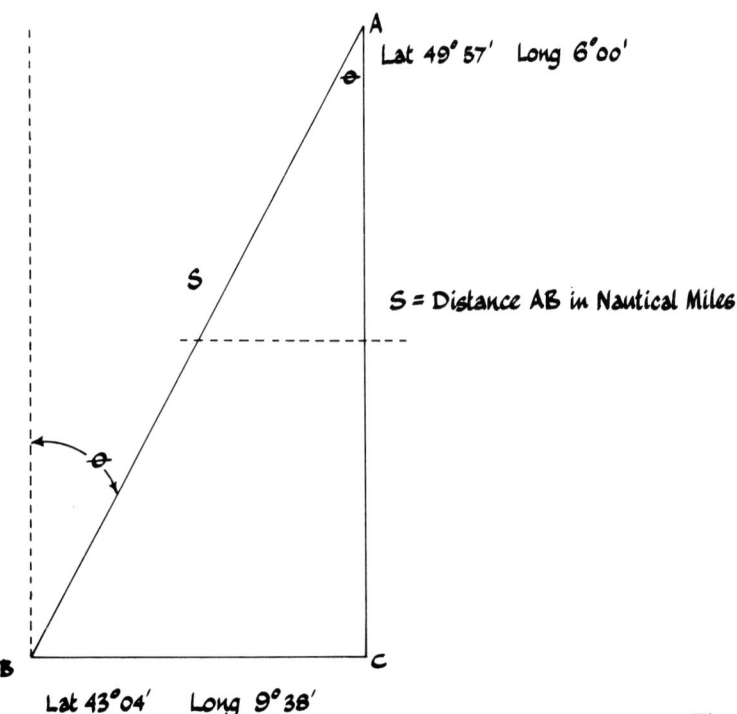

Fig. 4.12

Astro-Navigation with a Calculator

from which $\theta = 20°.135$.

The course sailing from A to B will be $20°.135 + 180° = 200°$.

and the distance $\quad S = \dfrac{6.883 \times 60}{\cos 20.135} = 439.86$ miles.

For practical purposes the ship should sail on a course $200°$ and sail 440 miles.

Books on navigation abound with examples worked out using rather complicated tables, but you can derive satisfaction by rapidly working out positions involving latitudes and longitudes, courses and distances run using plane right angled triangles, when distances are less than 600'.

4.15 Rhumb Lines and Courses using a Calculator

In the example of a yachtsman sailing in the Atlantic, Section 4.2, we found that his distance from the Cornish coast was 3191 nautical miles by great circle sailing. Suppose the yachtsman wished to sail on the somewhat longer course by rhumb line, to avoid having to change course several times. We can calculate what the steady rhumb line course should be using a calculator, without charts or tables. It is however advisable to check calculations using a Mercator chart (see Fig. 4.13).

From the spherical triangle relating to this example, and using the sine formula [3.7(7)] we have,

$$\sin C = \frac{\sin(\text{difference in Long})\cos\phi}{\sin(\text{distance})}$$

where C is the course at the start, at longitude $69°20'$W, latitude $26°40'$,

thus, $\quad \sin(\text{course at start}) = \dfrac{\sin 63°.583 \cos 50°.166}{\sin 53°.184}$

giving course at start $45°.77$.

If a great circle is followed, then course at the end of the trip will be given by,

$$\sin(\text{course at the end}) = \frac{\sin 63°.583 \cos 26°.66}{\sin 53°.184}$$

The calculator gives this course as 88.73 (see Fig. 4.13).

Thus for this great circle sailing the course changes gradually from $45°.77$ to $88°.73$.

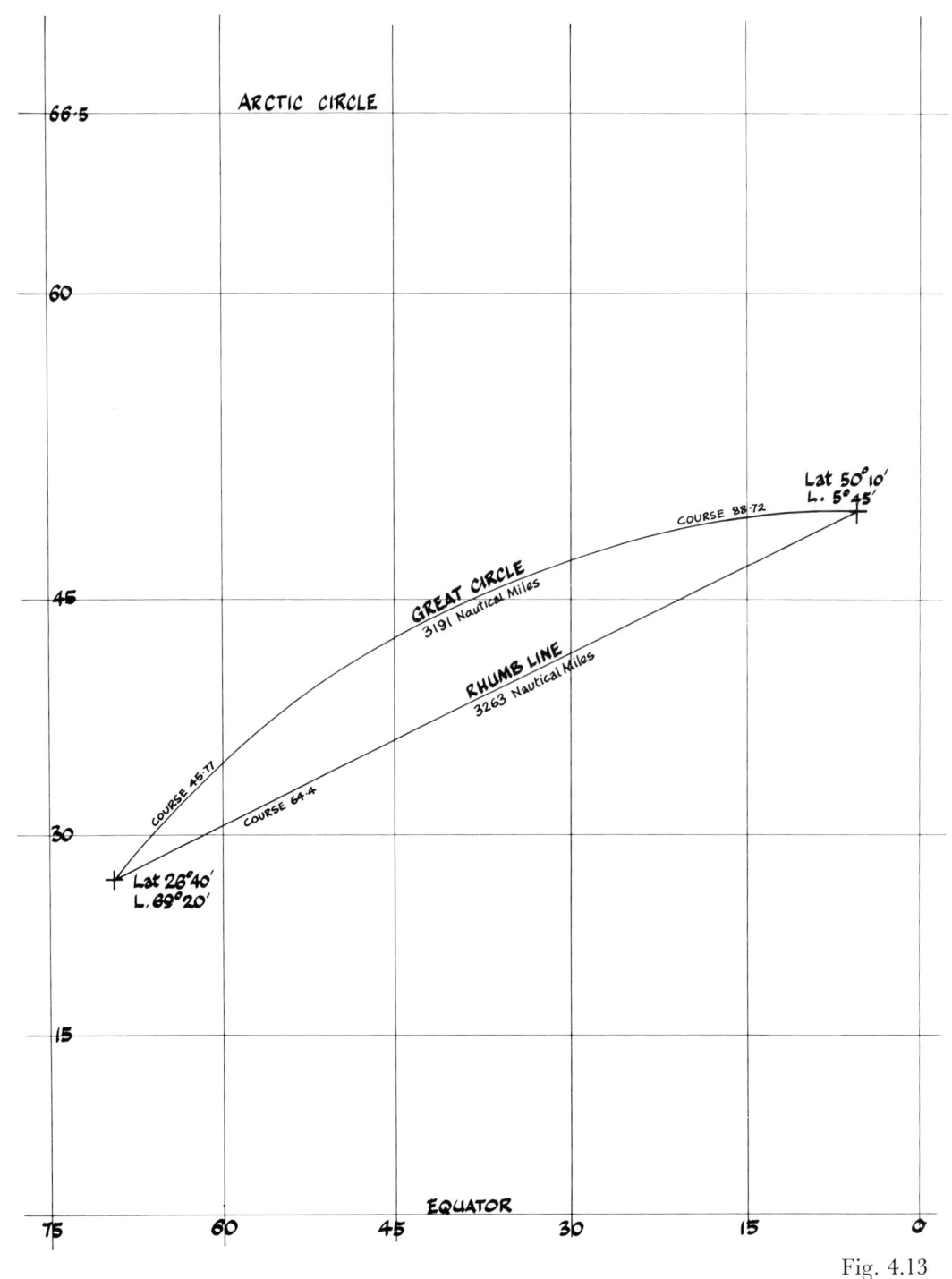

Positional Astronomy and Astro-Navigation Made Easy

Fig. 4.13

Astro-Navigation with a Calculator

A piece of cotton stretched between the point of departure (using a globe of the earth) and the point of arrival will illustrate this change of course which should take place in theory gradually, but in practice is carried out in several small changes.

We might expect a steady course to be somewhere between 45°.77 and 88°.73. Midway between these is about 67°, but the accurate steady course must be calculated as under.

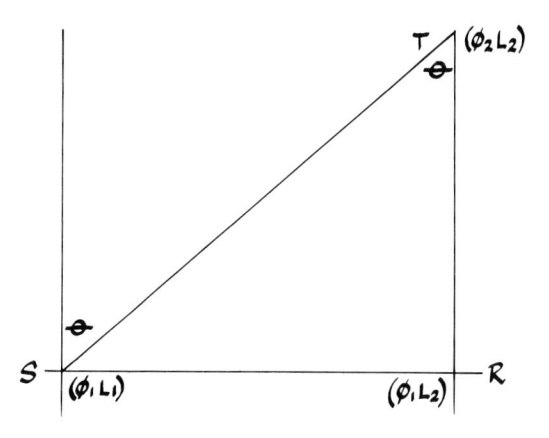

Fig. 4.14

4.16 The rhumb line *steady* course to be steered can be found graphically by plotting the two points (A and B) of departure and arrival on a Mercator chart and joining them by a straight line, and measuring the angle this rhumb line makes with the N–S longitude lines of the chart (see Fig. 4.14).

It is an interesting and useful exercise to *calculate* this course from the fact that a steady course on a Mercator chart is a straight line and that the tangent of the course, θ, required is basically given by the following ratio:

$$\tan \theta = \frac{\text{difference in longitudes} \ldots \text{(referred to the equator)}}{\text{difference in latitudes (after making proper allowance for the fact that the latitude scale on a Mercator chart is one of increasing proportions as we go from the equator towards the pole)}}$$

A study of a small scale Mercator chart will make this clear, and confirm that 1° of longitude at latitude ϕ will cover only $60 \cos \phi$ miles.

In a Mercator chart we have a longitude scale stretched at all latitudes by a factor of $\sec \phi$, so we have to stretch correspondingly the latitude scale of distances from the equator. This is not a simple uniform stretching but an ever increasing stretch as we go from the equator to the pole. At the pole the stretching becomes infinite or completely out of hand which explains why a Mercator chart cannot cope with polar regions.

Positional Astronomy and Astro-Navigation Made Easy

In Fig. 4.14, S and T represent two points on a Mercator chart. SR represents the difference in longitude, TR represents the difference in latitude, with due allowance made for the stretching of the scale referred to above. θ is the steady course.

Then

$$\tan(\text{course}) = \frac{SR}{TR}.$$

SR is simply $(L_2 - L_1)$ as S and R are on a Mercator chart, but TR has to be calculated, as follows.

4.17 In Fig. 4.15 $PBESW$ represents the Earth and B is a point in latitude ϕ. CE is the equator, R is the radius of the Earth.

In a Mercator projection chart each element of the Earth's surface, such as $Rd\phi$ subtended by a small change of latitude, $d\phi$ appears on a Mercator chart to cover a distance $R\,d\phi\,\sec\phi$.

At any point therefore in latitude ϕ' the total stretching from the equator, where $\phi = 0$ to ϕ' is the accumulation of all these stretched elements from E to B.

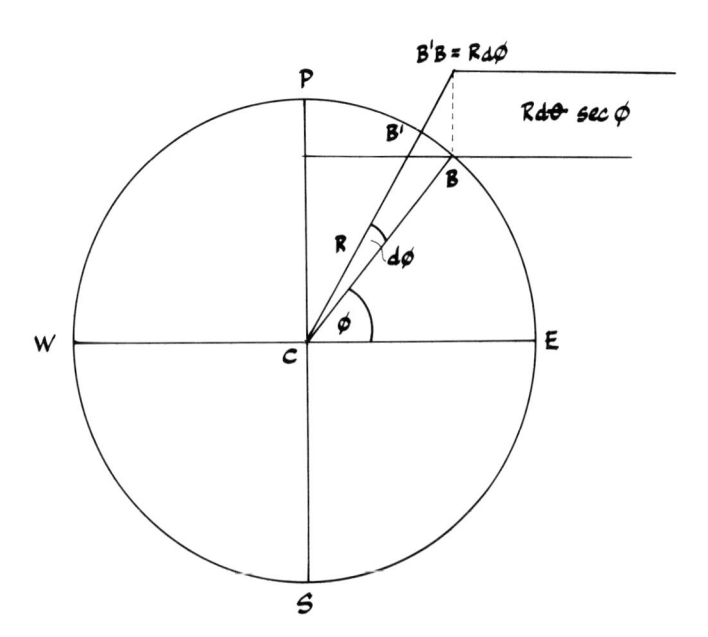

Fig. 4.15

Astro-Navigation with a Calculator

If y is the distance along a meridian from the equator *on the chart* from the equator, then $dy = R\,d\phi\,\sec\phi$ and to find the addition of all the elementary stretches we apply the integral calculus and write,

$$y = R\int_0^\phi \sec\phi\,d\phi.$$

This will be recognised as a standard form in calculus, and

$$y = R\log_e \tan\left(45+\frac{\phi}{2}\right).$$

We now have a means of calculating TR in Fig. 4.16 and so we are able to find the tangent of the course.

$$TR = R[\log_e \tan(45+\tfrac{1}{2}\phi_1) - \log_e(45+\tfrac{1}{2}\phi_2)]$$

and the course θ from Fig. 4.16 can now be obtained from

$$\tan\theta = R\left(\frac{L_2 - L_1}{\log_e \tan(45+\tfrac{1}{2}\phi_1) - \log_e \tan(45+\tfrac{1}{2}\phi_2)}\right).$$

In this formula L_1 and L_2 are expressed in *minutes* and R, the radius of the Earth, in nautical miles, which are equivalent to minutes.

Apply to our example in which $\phi = 50°10'$

and $\phi_2 = 26°40'$

$$\tan(\text{course}) = \frac{63°.583 \times 60}{3436 \log_e [\tan 70°.05 \div \tan 58°.33]} = 2.094$$

The calculator gives, the steady course $= 64°.47$ (calculators have a 'log$_e$' key usually marked l_n, and this makes light work of this calculation). The distance run on this course is given by,

$$\frac{\text{difference in Lat} \times 60}{\cos \text{rhumb line course}} = \frac{23°.5 \times 60}{\cos 64°.47}$$

$$= 3272 \text{ nautical miles}.$$

This is 81 miles longer than the great circle 3191 miles, calculated in Section 4.2.

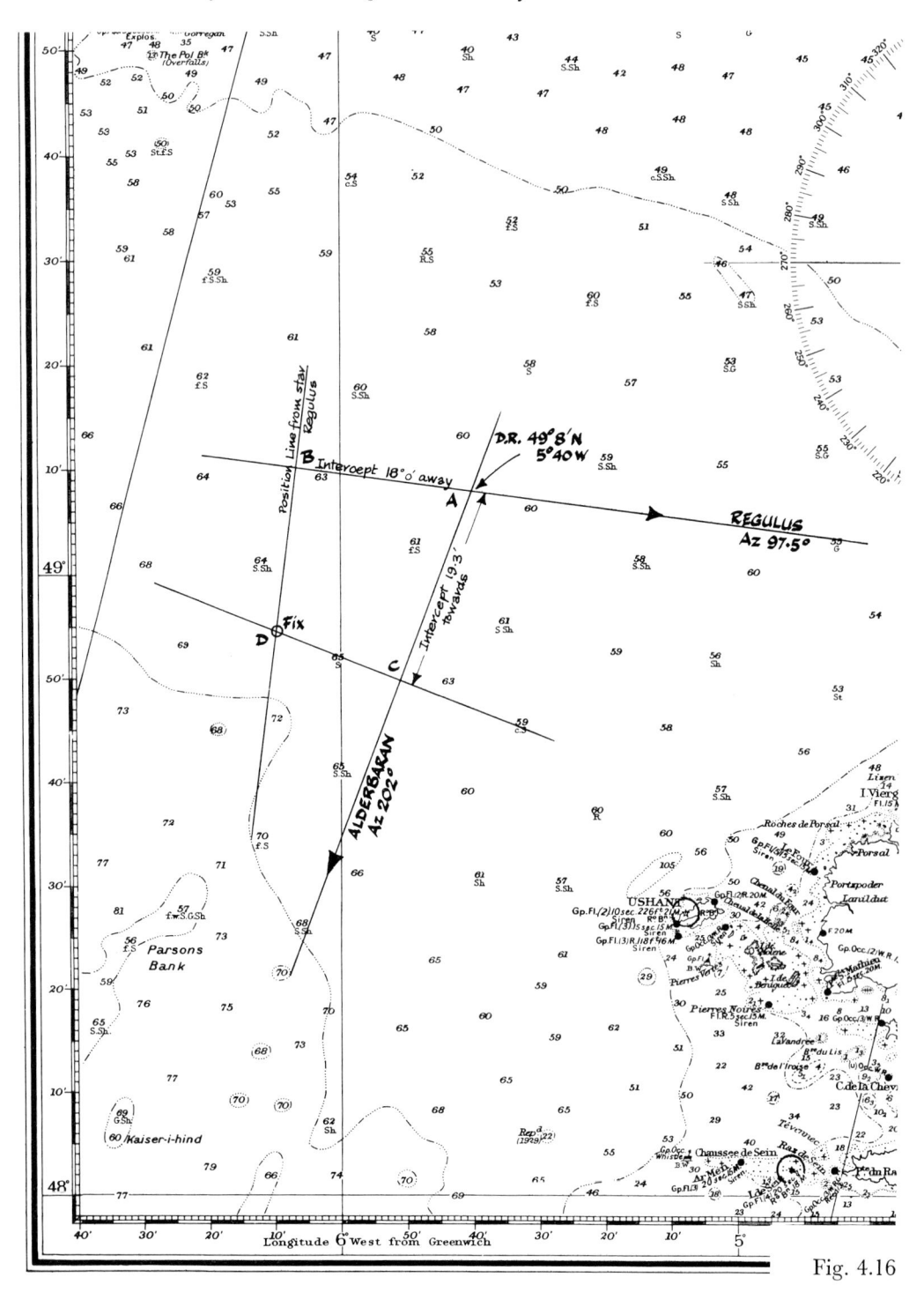

Positional Astronomy and Astro-Navigation Made Easy

Fig. 4.16

Astro-Navigation with a Calculator

4.18 A suggested method of recording and using data for a position line from a *star's* altitude, by sextant—and using a pocket calculator.

(1) Date (ascertain watch error from time signals). = 1976 Sept. 30

(2) Estimated position by DR (*A* on the chart). Lat ϕ 49° 08' N Long 5° 40' W°

(3) Star to be observed. = Aldebaran

(4) Index error of sextant. = Nil

(5) Observed *altitude* of star by sextant, *H*. = 56° 5.5'

(6) Exact time of observation (GMT). = 5 Hrs. 12 M 4 Sec

(7) (GHA) ♈ for the hour (6) from *tables* on day from *Nautical Almanac* = 84° 04.3'

   +increment for minutes and seconds from *yellow pages* = 3° 01.5'

   (GHA) ♈ = 87° 05.8'

   +SHA for star (*Nautical Almanac*) = 291° 20.9'

   (GHA) ♈ +(SHA)* = 378° 26.7'   (GHA) star = 18° 26.7'

   −Long W° or +Long E° = −5° 40'

   LHA star. = 12° 46.7'

(8) Declination of star (δ) from *Nautical Almanac*. 16° 27.8' = 16°.463

(9) Calculated H_c from ϕ, δ, and LHA from (2), (8) (7), and relation:
   $\sin H_c = \sin \phi \sin \delta + \cos \phi \cos \delta \cos(\text{LHA})$
   $\sin H_c = \sin 49°.133 \sin 16°.463 + \cos 49°.133 \cos 16°.463 \cos 12°.778$
   by calculator. Whence $H_c = $ 55°.716 = 55° 43'

(10) Sextant reading (5). = 56° 5.5'

   (a) Index error of sextant. = Nil

   (b) Correction for dip—always *subtracted*.
      = $1.92' \sqrt{M}$ where *M* = height above sea level in metres (2 m)
      or $1.06' \sqrt{h}$ where *h* = height above sea level in feet. = −2.7'

   (c) Correction for refraction from table in *Nautical Almanac* (corresponding to altitude (10). *Subtract*. = −0.5'

   Corrected H_0. = 56° 02.3'

105

(11) Azimuth of star from DR position, given by:
$$\sin(Az) = \frac{\sin(LHA)\cos\delta}{\cos H_c}$$
giving $(Az) \pm 180$ = 202°.12

(12) $H_0 - H_c$ (from (9) and (10)) in ′ of arc = 56° 02′.3 − 55° 43′

This is the intercept in minutes of arc or nautical miles, and is *towards* the star if H_0 is greater than H_c. (*This is AB on the chart*, Fig. 4.16).

= 19′.3

(13) The results of (11) and (12) are sufficient for drawing the position line.

(14) *Repeat procedure for another star or body well separated from the first to obtain a fix.*

Star (2)

4.19 A suggested method of recording and using data for a position line from a *star's* altitude, by sextant—and using a pocket calculator.

(1) Date (ascertain watch error from time signals). = 1976 Sept. 30

(2) Estimated position by DR. Lat ϕ 49°.08 Long 5°.40 W°

(3) Star to be observed. Regulus

(4) Index error of sextant. = nil

(5) Observed *altitude* of star by sextant, H (uncorrected). = 22° 12′.6

(6) Exact time of observation (GMT). = 5 Hrs. 15 M 9 Sec
(7) (GHA) ♈ for the hour (6) from *tables* on day (*Nautical Almanac*) = 84° 04′.3
+increment for minutes and seconds from *yellow pages* = 3° 47′.8

(GHA) ♈ = 87° 52′.1
+SHA for star (*Nautical Almanac*) = 208° 13′.1
(GHA) ♈ + (SHA)* = 296° 05′.2   (GHA) star = 296° 05′.2
− Long W° or + Long E° = − 5° 40′
LHA star. = 290° 25′.2

(8) Declination of star (δ) from *Nautical Almanac*. (N) = 12° 04′.9

(9) Calculated H_c from ϕ, δ, and LHA from (2), (8) and (7), and relation:
$\sin H_c = \sin\phi \sin\delta + \cos\phi \cos\delta \cos(LHA)$
$\sin H_c = \sin 49°.133 \sin 12°.082 + \cos 49°.133 \cos 12°.082 \cos 290.42$
by calculator. Whence $H_c = 22°.4277$ = 22° 25′.7

Astro-Navigation with a Calculator

(10) Sextant reading (5). $= 22°12.6'$

   (a) Index error of sextant. $= nil$

   (b) Correction for dip—always subtracted.
$= 1.92' \sqrt{M}$ where $M =$ height above sea level in metres. **(2 M)**
or $1.06' \sqrt{h}$ where $h =$ height above sea level in feet. $= -2.7'$

   (c) Correction for refraction from table in *Nautical Almanac* (corresponding to altitude (10)). Subtract. $= -2.2'$

   Corrected H_0. $= 22°07.7'$

(11) Azimuth of star from DR position, given by:
$$\sin(Az) = \frac{\sin(LHA)\cos \delta}{\cos H_c} \quad \overset{-82°47}{\text{giving}} \quad (Az) \pm 180.$$
$= 97.5°$ from N

(12) $H_0 - H_c$ (from (9) and (10)) in ' of arc. $= 22°25.7' - 22°07.7'$

This is the intercept in minutes of arc or nautical miles, and is *towards* the star if H_0 is greater than H_c.
$= 18'$ AWAY.

(13) The results of (11) and (12) are sufficient for drawing the position line.

We now have two position lines intersecting at D, which gives the ship's position (see Fig. 4.16).

4.20 A suggested method of recording data for a position line from the sextant altitude of the *Sun*, *Moon* or a *planet*—using a pocket calculator.

(1) Date (ascertain watch error from time signals). ....

(2) Estimated position by DR.   Lat ϕ ...... Long...... W° or E°

(3) Which body, Sun, Moon or Planet. ....

(4) Index error of sextant. $= $ ....

(5) Observed *altitude* of body by sextant, H. $= $ ....

(6) Exact time of observation (GMT). $= $ ...Hrs......M...Sec

(7) For Sun, Moon or planets, take the GHA from the *Nautical Almanac* for the hour. $= $ ....

Increments for minutes and seconds. $= $ ....

107

Positional Astronomy and Astro-Navigation Made Easy

    v correction to be applied to Moon or planet (see *Nautical Almanac*) (this is a simple interpolation between hours). = ...

    GHA of Sun, Moon or planet = ...

    − Long W°   or   +Long. E° = ...

    Local hour angle of Sun, Moon or planet (LHA). = ...

(8) Declination of Sun, Moon or planet at hour = ...

    ±d correction as the declination changes hourly (especially important for Moon observations) Correction + or − by inspection. = ...

(9) Calculated altitude H_c from ϕ, δ, and LHA from (2), (8) and (7). From $\sin H_c =$ $\sin \phi \sin \delta + \cos \phi \cos \delta \cos(\text{LHA})$.
i.e., $\sin H_c = \sin...... \sin...... + \cos...... \cos......$
$\cos......$ By calculator. H_c = ...

(10) Corrections for sextant altitude (5). Reading H = ...

    Index error ± = ...

    +semi-diameter of Sun or Moon (using lower limb) from *Nautical Almanac, minus* SO using upper limb (approx. 16′)
(not for bubble sextant) = ...

    Dip always subtracted.
        $1.92′ \sqrt{M}$ where M = number of metres above sea level
        $1.06′ \sqrt{h}$ where h = number of feet above sea level
of sextant. = ...

    Refraction of atmosphere from table in *Nautical Almanac (subtract)* = ...

    For the *Moon* a large parallax correction is necessary, found from: HP $\cos H_c$ (HP from the daily tables in *Nautical Almanac*) (always added). = ...

    Corrected H_c = ...

(11) Azimuth of body from the DR position, given by:
$$\sin(\text{Az}) = \frac{\sin(\text{LHA})\cos \delta}{\cos H_c} \text{ giving (Az)}$$
= ....

(12) $H_o - H_c$ (from (9) and (10)) in minutes of arc = ...

This is the *intercept* in minutes of arc or nautical miles and is *towards* the Sun, Moon or planet if H_o is greater than H_c, in the direction of azimuth from DR position.

(13) The results of (11) and (12) are sufficient for drawing the position line.

(14) *Repeat procedure for another star or body well separated from the first to obtain a fix.*

4.21 The Use of the Calculator to get the Azimuth of a Body and thence to find the Compass Error

It is important to know just how the compass behaves under various conditions of the boat's head. For example, iron and steel structures, e.g., engines, will influence the compass in different directions depending on the boat's magnetic course. The difference between a boat's proper magnetic course (i.e., assuming the boat to contain no magnetic material) and the course shown on the compass in an actual boat containing iron and steel is the *deviation*.

This is quite separate from the *variation* which is the difference between the true geographical N–S meridian and the magnetic N–S at the locality of the boat. This *varies* from place to place. Deviation depends on the boat's heading.

The compass error is a combination of deviation and variation. The variation is given on the chart. The deviation is found by swinging the boat, and bringing her head so that the compass reads in turn 0°, 45°, 90°, 135°, 180°, 225°, 270°, 315° and 360° on each head. The exact bearing by compass of a distant object of known magnetic or true bearing is noted. Make a table:

Ship's Head	Dev.
—	—

A very useful method is to use the azimuth of the Sun or a star as the distant object of known bearing. This azimuth is given in Table 5.12 or it can be calculated as we have been doing using the relation [3.7(1)] or [3.7(5)],

$$\sin(\text{Az}) = \frac{\sin(\text{HA}) \cos \delta}{\cos(\text{alt})}.$$

(Using HA from *Nautical Almanac*, and altitude by sextant direct, or by $\sin(\text{alt}) = \sin \phi \sin \delta + \cos \phi \cos \delta \cos(\text{HA})$) as was done in [4.5(8)] [4.6(8)], [4.18(11)]. See also Section 3.10.

There are several 'rules' for determining just how to find the compass error from the variation and deviation. They are complicated, easily forgotten and unsatisfactory. The best rule is *draw a diagram* as in Fig. 4.17.

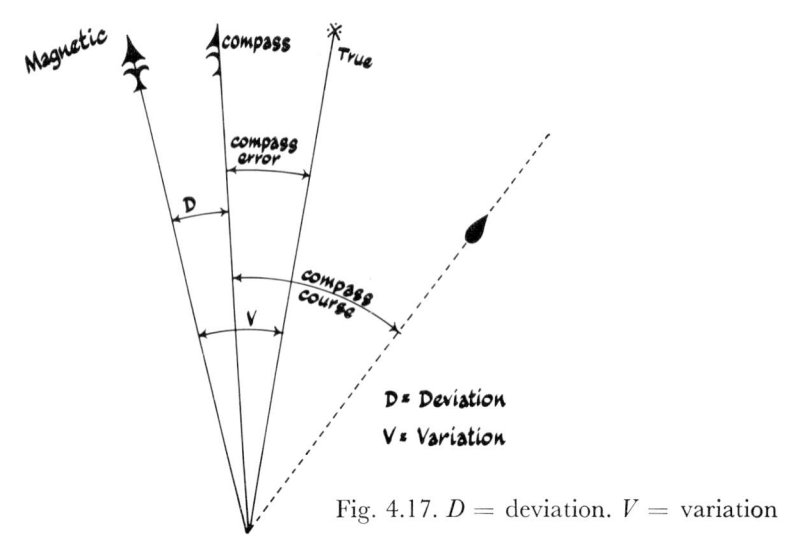

Positional Astronomy and Astro-Navigation Made Easy

Fig. 4.17. D = deviation. V = variation

The advantage of using the Sun is that you do not rely on a land object. You can get the azimuth accurately and the difference between the bearing of the object (true) and compass bearing is the compass error. From this, apply the variation to get the *deviation*. Deviations are generally small in boats containing small amounts of iron, and can be corrected by expert placing of small compensating magnets.

4.22 The Prime Vertical Altitude

If we know our latitude ϕ and the declination δ of a body we can with a little patience find due east or due west by using the relation [3.7(4)], $\sin \delta = \sin \phi \sin(\text{alt}) + \cos \phi \cos(\text{alt}) \cos(\text{Az})$ and putting $(\text{Az}) = 90$ or 180 so that $\cos(\text{Az}) = 0$.

Then when
$$\sin(\text{alt}) = \frac{\sin \delta}{\sin \phi}$$

then a celestial body is due east or due west. It is on the 'prime vertical'.

This fact can also be made use of to check the compass as the following example will show.

Suppose the Sun's declination is $17°.43$ N and the latitude of the observer is known to be $49°.7$ N. Then,

$$\sin(\text{alt}) = \frac{\sin 17°.43}{\sin 49°.7}$$

giving the altitude as $23°.126$ or $23°\ 7'.56$.

Astro-Navigation with a Calculator

Set and keep the sextant for this reading and observe the Sun for a few minutes from the time the Sun has an altitude of say 22° (approx) until it reaches the 'set value'. At this instant note the compass bearing of the Sun and compare with the true bearing of 90°. For an observation when the body is to the west of the meridian, and the altitude is therefore falling, the observer would start his time of observing a few minutes before the altitude falls to the pre-set altitude on the sextant. The bearing will then be 270° (true). The compass error (deviation+variation) can thus be ascertained. When a body is due east or west its altitude changes rapidly with time and this helps in making an accurate observation of its position.

4.23 **A Simple Navigational Puzzle**

A navigator in a jet aircraft when flying over the North Pole in late summer is heading for London (lat 51°, long 0°). He gets a bubble sextant sight of the Sun at 1000 GMT and finds that it is exactly 15°. What was the date?

He keeps on a course for London and 2 hours and 5 minutes later again observes the Sun and finds it to have an altitude of 36°30′. How far had he travelled, and at what speed?

What is his estimated time of arrival in London? Assume constant speed.

4.24 *Solution:* The Sun has an almost constant altitude above the horizon at the pole throughout the summer's day and night. This altitude is the same as the Sun's declination on that date. The *Nautical Almanac* puts that date as 12th August. (It could have been the 1st May, but this is not late summer.)

At 1205 GMT on this date the Sun is on meridian passage. Meridian passage is at GMT+equation of time (5 m) on this date. The altitude of the Sun is given by

$$90 - \phi + 15 = 36°30'$$

where ϕ is the latitude at 1205 GMT.

So that the aircraft's latitude, ϕ, is 68°30′.

The aircraft must have travelled from the North Pole a distance of

$$(90 - 60°30') \times 60 \text{ nautical miles in } 2^h 5^m$$

a distance of 1290 nautical miles in 2.083 3 hours.

Speed of aircraft = 619.21 knots.

London is due South at a distance of $(68°30'-51) \times 60 = 1050$ nautical miles.

Time to travel this distance is $\dfrac{1050}{619.21} = 1.695$ hours or $1^h 42^m$.

Expected time of arrival $= 12.05 + 1^h 42^m = 13.47$ GMT.

By travelling along the Greenwich meridian we need not be concerned with changes in longitude. On this flight there was a slight tendency for the aircraft to drift off course to the west. Can you account for this?

4.25 Your latitude can be found by a simple calculation using only two altitudes, two Azimuths and one unknown star!

Example: A Star is observed to have altitude 18° bearing 41°. Some time later the *same* star observed from the same position has altitude 72° bearing 109°.

Suppose the declination to be δ, then using relation [3.7(4)] and the observed altitudes and azimuths we have,

$$\sin \delta = \sin \phi \sin 18 + \cos \phi \cos 18 \cos 41 \qquad [4.24(1)]$$

and also,

$$\sin \delta = \sin \phi \sin 72 + \cos \phi \cos 72 \cos 109 \qquad [4.24(1)]$$

we can eliminate δ thus:

$$\sin \phi \sin 18 + \cos \phi \cos 18 \cos 41$$
$$= \sin \phi \sin 72 + \cos \phi \cos 72 \cos 109$$

or,
$$\frac{\sin \phi}{\cos \phi} = \frac{\cos 72 \cos 109 - \cos 18 \cos 41}{\sin 18 - \sin 72}$$

a simple calculator exercise giving,

$$\tan \phi = 1.2746$$
$$\phi = 51°.8848$$

5
Alt-Azimuth Curves Plotted for Various Planispheres using the Calculator

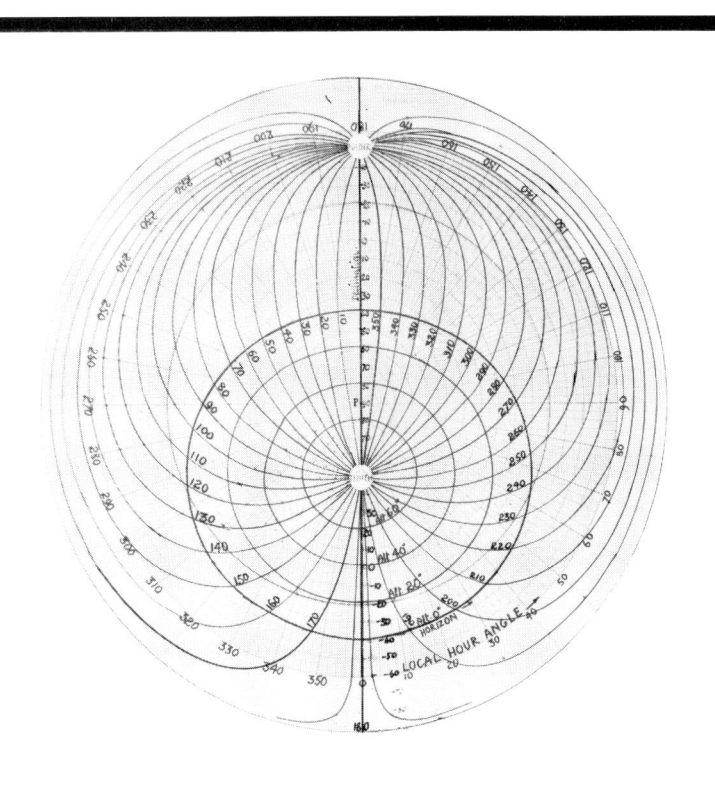

5.1

Star maps such as those depicted on a planisphere, or on star charts of various types with coordinates for RA and declination, have a serious limitation in that they do not indicate the all-important position of a heavenly body in terms of its *altitude* and *azimuth*—the natural earthly coordinates of a star at a particular time.

Air navigators are familiar with star maps that have a graticule of azimuth lines, or altitude curves to provide this information, and it occurred to the author that with the help of a calculator it might be interesting, instructive and within the capability of 'A' level students of mathematics to find the coordinates of both azimuth curves and altitude curves for use on a planisphere. This seemed to be a suitable exercise for the pocket scientific calculator, and so it proved to be.

In order to plot these curves for use on a planisphere having *polar* coordinates, we require sets of data connecting δ, HA and Az, but keeping ϕ, the latitude of the place, constant.

Then for a particular azimuth we calculate the δ for various HA's from 0 to 180 and so produce a curve of equal azimuths for this declination. The only relation that appears to be satisfactory for this purpose is the four part formula [3.7(8)] and Section 3.12.

$$\cot(\text{Az}) = \frac{\cos\phi \tan\delta - \sin\phi \cos(\text{HA})}{\sin(\text{HA})}$$

or,
$$\tan\delta = \frac{\sin\phi \cos(\text{HA}) + \sin(\text{HA}) \cot(\text{Az})}{\cos\phi}.$$

As an example we consider the equal Az curve of say 60° and HA's from 10 to 180 with $\phi = 51$. i.e., we keep Az constant at 60°, $\phi = 51$ and we vary the HA.

For (HA) = 10
$$\tan\delta = \frac{\sin 51° \cos 10° + \sin 10° \cot 60°}{\cos 51}$$

$$\delta = 53°.98.$$

The calculator results δ are then as follows:

for constant Az

HA	δ
10°	53°.98
20°	55°.85
30°	56°.80
40°	56°.93
60°	54°.69
80°	48°.19

etc. as shown in the Table 5.2 (51°), column 6. These points for δ all lie on the (Az) = 60° curve.

Altitude and Azimuth Lines

5.2 Tables 5.1–5.5 (40°, 51°, 53°, 55° and 57°) give the results for all azimuths from 10° to 180° for latitudes 40°, 51°, 53°, 55° and 57°, each single calculation requiring less than one minute to perform.

The tables were compiled using the hand held calculator, but were checked by a digital computer by the courtesy and cooperation of the staff of the Salisbury College of Technology who used the computer terminal facilities of the college. It was gratifying to find that humble efforts on a calculator, costing a few pounds produced results identical with those produced by a computer costing thousands!

The results for the latitudes north of the equator are equally applicable to latitudes south of the equator. Results can be calculated and curves drawn for any latitude, and it is satisfying to have a set of results for your own particular latitude, e.g., lat 40°N for the middle latitudes of the United States, and lat 51°N for the South of England. The results for lat +40° are valid for −40° in the southern hemisphere (see Section 5.6).

5.3 The results for latitude 51° were plotted on 'centimetric' polar graph paper (r, θ) with δ along the radius vector and the HA for θ shown in Fig. 5.1.

5.4 It is of interest to note that the relation does not recognise any horizon barrier since the altitude does not enter into the expression. This means that each azimuth curve begins at the zenith point of the celestial sphere or on a star map graticule and continues on its way past the horizon and through the nadir point and finally completing the circle back to the zenith, thus making the interesting pattern of curves shown in Fig. 5.1 which gives, as a matter for interest, the azimuth of celestial objects even though they are well below the horizon or on the other side of the world!

5.5 ## Altitude Curves (or Almucantars)

In order to identify a star or find out where to look for it, we need in addition to the azimuth, the star's *altitude*. Here we can use the calculator to plot curves of equal altitude, and in this case we use relation [3.3(2)]

$$\sin(\text{alt}) = \sin \delta \sin \phi + \cos \delta \cos \phi \cos(\text{HA})$$

transformed to

$$\cos(\text{HA}) = \frac{\sin(\text{alt}) - \sin \phi \sin \delta}{\cos \phi \cos \delta}$$

(see Tables 5.5–5.11).

Positional Astronomy and Astro-Navigation Made Easy

5.6(40°) Table of coordinates of points for plotting azimuth curves relating to latitude 40°N (see Section 5.2) for use on star charts of various types, Polar, Cartesian, Mercator or Stereographic projection. The relation used is

$$\tan \delta = \frac{\sin \phi \cos(\mathrm{HA}) + \sin(\mathrm{HA}) \cot(\mathrm{Az})}{\cos \phi}$$

where $\phi = 40°$ (all figures are in degrees).

Table 5.1

(HA) \ δ (Az)→	10	20	30	40	50	60	70	80	90
10	64.66	55.39	50.64	47.64	45.47	43.75	42.27	40.90	39.57
20	73.24	63.61	57.37	52.87	49.31	46.30	43.56	40.93	38.26
30	77.28	68.36	61.70	56.39	51.88	47.82	43.96	40.09	36.01
40	79.51	71.26	64.50	58.67	53.41	48.42	43.48	38.33	32.73
50	80.85	73.08	66.24	59.99	54.04	48.16	42.09	35.59	28.34
60	81.67	74.16	67.19	60.49	53.84	47.00	39.73	31.75	22.76
70	82.14	74.71	67.48	60.24	52.78	44.86	36.26	26.72	16.01
80	82.34	74.79	67.14	59.20	50.76	41.60	31.53	20.42	8.29
90	82.31	74.42	66.14	57.27	47.61	37.00	25.41	12.96	.00
100	82.03	73.55	64.33	54.20	43.02	30.82	17.86	4.63	−8.29
110	81.47	72.03	61.45	49.60	36.59	22.84	9.06	−4.04	−16.01
120	80.53	69.58	56.98	42.85	27.88	13.12	−.46	−12.42	−22.76
130	78.97	65.64	50.02	33.12	16.69	2.18	−9.95	−19.95	−28.34
140	76.34	58.97	39.03	19.66	3.51	−9.00	−18.64	−26.33	−32.73
150	71.42	46.85	21.99	2.93	−10.15	−19.28	−26.06	−31.45	−36.01
160	60.16	23.66	−.87	−14.38	−22.48	−27.96	−32.05	−35.37	−38.26
170	24.67	−11.51	−23.45	−29.08	−32.46	−34.82	−36.64	−38.18	−39.57
180	−40.00	−40.00	−40.00	−40.00	−40.00	−40.00	−40.00	−40.00	−40.00

(HA) \ δ (Az)→	100	110	120	130	140	150	160	170	180
10	38.18	36.64	34.82	32.46	29.08	23.45	11.51	−24.67	−90.00
20	35.37	32.05	27.96	22.48	14.38	.87	−23.66	−60.16	−90.00
30	31.45	26.06	19.28	10.15	−2.93	−21.99	−46.85	−71.42	−90.00
40	26.33	18.64	9.00	−3.51	−19.66	−39.03	−58.97	−76.34	−90.00
50	19.95	9.95	−2.18	−16.69	−33.12	−50.02	−65.64	−78.97	−90.00
60	12.42	.46	−13.12	−27.88	−42.85	−56.98	−69.58	−80.53	−90.00
70	4.04	−9.06	−22.84	−36.59	−49.60	−61.45	−72.03	−81.47	−90.00
80	−4.63	−17.86	−30.82	−43.02	−54.20	−64.33	−73.55	−82.03	−90.00
90	−12.96	−25.41	−37.00	−47.61	−57.27	−66.14	−74.42	−82.31	−90.00
100	−20.42	−31.53	−41.60	−50.76	−59.20	−67.14	−74.79	−82.34	−90.00
110	−26.72	−36.26	−44.86	−52.78	−60.24	−67.48	−74.71	−82.14	−90.00
120	−31.75	−39.73	−47.00	−53.84	−60.49	−67.19	−74.16	−81.67	−90.00
130	−35.59	−42.09	−48.16	−54.04	−59.99	−66.24	−73.08	−80.85	−90.00
140	−38.33	−43.48	−48.42	−53.41	−58.67	−64.50	−71.26	−79.51	−90.00
150	−40.09	−43.96	−47.82	−51.88	−56.39	−61.70	−68.36	−77.28	−90.00
160	−40.93	−43.56	−46.30	−49.31	−52.87	−57.37	−63.61	−73.24	−90.00
170	−40.90	−42.27	−43.75	−45.47	−47.64	−50.64	−55.39	−64.66	−90.00
180	−40.00	−40.00	−40.00	−40.00	−40.00	−40.00	−40.00	−40.00	

Altitude and Azimuth Lines

Table 5.2

5.6(51°) Table of coordinates of points for plotting azimuth curves relating to latitude 51°N for use on star charts of various types, Polar, Cartesian, Mercator or Stereographic projection. The relation used is

$$\tan \delta = \frac{\sin \phi \cos(\text{HA}) + \sin(\text{HA}) \cot(\text{Az})}{\cos \phi}$$

δ (HA) \ (Az)	10	20	30	40	50	60	70	80	90
10	70.22	63.14	59.45	57.09	55.36	53.98	52.78	51.67	50.57
20	76.74	69.35	64.56	61.05	58.26	55.85	53.64	51.48	49.25
30	79.83	72.91	67.76	63.62	60.06	56.80	53.65	50.42	46.92
40	81.56	75.08	69.78	65.19	60.99	56.93	52.81	48.39	43.41
50	82.60	76.41	70.99	65.98	61.15	56.25	51.04	45.24	38.44
60	83.23	77.19	71.57	66.11	60.56	54.69	48.20	40.70	31.69
70	83.58	77.54	71.61	65.57	59.17	52.10	44.00	34.44	22.90
80	83.72	77.51	71.12	64.32	56.79	48.19	38.10	26.12	12.10
90	83.67	77.10	70.03	62.16	53.13	42.53	30.04	15.65	.00
100	83.41	76.24	68.17	58.79	47.69	34.57	19.55	3.52	−12.10
110	82.92	74.80	65.20	53.62	39.71	23.74	6.91	−9.04	−22.90
120	82.08	72.46	60.48	45.64	28.25	10.04	−6.65	−20.55	−31.69
130	80.70	68.59	52.74	33.30	12.82	−5.20	−19.33	−30.08	−38.44
140	78.34	61.74	39.46	15.18	−5.08	−19.61	−29.87	−37.45	−43.41
150	73.78	48.07	17.05	−6.99	−21.94	−31.41	−37.96	−42.90	−46.92
160	62.51	18.41	−12.36	−27.15	−35.16	−40.25	−43.91	−46.79	−49.25
170	19.23	−24.61	−36.44	−41.58	−44.56	−46.58	−48.13	−49.42	−50.57
180	−51.00	−51.00	−51.00	−51.00	−51.00	−51.00	−51.00	−51.00	−51.00

δ (HA) \ (Az)	100	110	120	130	140	150	160	170	180
10	49.42	48.13	46.58	44.56	41.58	36.44	24.61	−19.23	−90.00
20	46.79	43.91	40.25	35.16	27.15	12.36	−18.41	−62.51	−90.00
30	42.90	37.96	31.41	21.94	6.99	−17.05	−48.07	−73.78	−90.00
40	37.45	29.87	19.61	5.08	−15.18	−39.46	−61.74	−78.34	−90.00
50	30.08	19.33	5.20	−12.82	−33.30	−52.74	−68.59	−80.70	−90.00
60	20.55	6.65	−10.04	−28.25	−45.64	−60.48	−72.46	−82.08	−90.00
70	9.04	−6.91	−23.74	−39.71	−53.62	−65.20	−74.80	−82.92	−90.00
80	−3.52	−19.55	−34.57	−47.69	−58.79	−68.17	−76.24	−83.41	−90.00
90	−15.65	−30.04	−42.53	−53.13	−62.16	−70.03	−77.10	−83.67	−90.00
100	−26.12	−38.10	−48.19	−56.79	−64.32	−71.12	−77.51	−83.72	−90.00
110	−34.44	−44.00	−52.10	−59.17	−65.57	−71.61	−77.54	−83.58	−90.00
120	−40.70	−48.20	−54.69	−60.56	−66.11	−71.57	−77.19	−83.23	−90.00
130	−45.24	−51.04	−56.25	−61.15	−65.98	−70.99	−76.41	−82.60	−90.00
140	−48.39	−52.81	−56.93	−60.99	−65.19	−69.78	−75.08	−81.56	−90.00
150	−50.42	−53.65	−56.80	−60.06	−63.62	−67.76	−72.91	−79.83	−90.00
160	−51.48	−53.64	−55.85	−58.26	−61.05	−64.56	−69.35	−76.74	−90.00
170	−51.67	−52.78	−53.98	−55.36	−57.09	−59.45	−63.14	−70.22	−90.00
180	−51.00	−51.00	−51.00	−51.00	−51.00	−51.00	−51.00	−51.00	

Table 5.3

5.6(53°) Table of coordinates of points for plotting azimuth curves relating to latitude 53°N. The relation used is

$$\tan \delta = \frac{\sin \phi \cos(\mathrm{HA}) + \sin(\mathrm{HA}) \cot(\mathrm{Az})}{\cos \phi}$$

(HA) \ δ(Az) →	10	20	30	40	50	60	70	80	90
10	71.23	64.53	61.04	58.79	57.15	55.84	54.69	53.63	52.58
20	77.39	70.40	65.86	62.54	59.88	57.59	55.48	53.41	51.27
30	80.32	73.75	68.88	64.95	61.56	58.45	55.44	52.34	48.97
40	81.95	75.80	70.77	66.41	62.40	58.52	54.57	50.31	45.47
50	82.94	77.05	71.89	67.12	62.50	57.80	52.78	47.14	40.46
60	83.53	77.78	72.42	67.20	61.88	56.21	49.89	42.53	33.57
70	83.87	78.10	72.43	66.63	60.45	53.58	45.63	36.10	24.41
80	84.00	78.05	71.93	65.36	58.05	49.61	39.56	27.43	12.98
90	83.94	77.64	70.84	63.21	54.35	43.81	31.17	16.33	.00
100	83.69	76.81	68.99	59.82	48.81	35.54	20.06	3.33	−12.98
110	83.21	75.39	66.04	54.60	40.57	24.11	6.53	−10.12	−24.41
120	82.40	73.09	61.33	46.44	28.54	9.50	−7.96	−22.28	−33.57
130	81.07	69.28	53.51	33.58	12.14	−6.74	−21.29	−32.15	−40.46
140	78.78	62.46	39.81	14.38	−6.86	−21.80	−32.12	−39.63	−45.47
150	74.32	48.58	16.16	−9.04	−24.33	−33.81	−40.26	−45.08	−48.97
160	63.16	17.45	−14.72	−29.67	−37.60	−42.58	−46.13	−48.91	−51.27
170	18.24	−27.21	−38.91	−43.92	−46.80	−48.75	−50.24	−51.47	−52.58
180	−53.00	−53.00	−53.00	−53.00	−53.00	−53.00	−53.00	−53.00	−53.00

(HA) \ δ(Az) →	100	110	120	130	140	150	160	170	180
10	51.47	50.24	48.75	46.80	43.92	38.91	27.21	−18.24	−90.00
20	48.91	46.13	42.58	37.60	29.67	14.72	−17.45	−63.16	−90.00
30	45.08	40.26	33.81	24.33	9.04	−16.16	−48.58	−74.32	−90.00
40	39.63	32.12	21.80	6.86	−14.38	−39.81	−62.46	−78.78	−90.00
50	32.15	21.29	6.74	−12.14	−33.58	−53.51	−69.28	−81.07	−90.00
60	22.28	7.96	−9.50	−28.54	−46.44	−61.33	−73.09	−82.40	−90.00
70	10.12	−6.53	−24.11	−40.57	−54.60	−66.04	−75.39	−83.21	−90.00
80	−3.33	−20.06	−35.54	−48.81	−59.82	−68.99	−76.81	−83.69	−90.00
90	−16.33	−31.17	−43.81	−54.35	−63.21	−70.84	−77.64	−83.94	−90.00
100	−27.43	−39.56	−49.61	−58.05	−65.36	−71.93	−78.05	−84.00	−90.00
110	−36.10	−45.63	−53.58	−60.45	−66.63	−72.43	−78.10	−83.87	−90.00
120	−42.53	−49.89	−56.21	−61.88	−67.20	−72.42	−77.78	−83.53	−90.00
130	−47.14	−52.78	−57.80	−62.50	−67.12	−71.89	−77.05	−82.94	−90.00
140	−50.31	−54.57	−58.52	−62.40	−66.41	−70.77	−75.80	−81.95	−90.00
150	−52.34	−55.44	−58.45	−61.56	−64.95	−68.88	−73.75	−80.32	−90.00
160	−53.41	−55.48	−57.59	−59.88	−62.54	−65.86	−70.40	−77.39	−90.00
170	−53.63	−54.69	−55.84	−57.15	−58.79	−61.04	−64.53	−71.23	−90.00
180	−53.00	−53.00	−53.00	−53.00	−53.00	−53.00	−53.00	−53.00	

Altitude and Azimuth Lines

Table 5.4

5.6(55°) Table of coordinates of points for plotting azimuth curves relating to latitude 55°N. The relation used is

$$\tan \delta = \frac{\sin \phi \cos(\text{HA}) + \sin(\text{HA}) \cot(\text{Az})}{\cos \phi}$$

(HA) \ (Az) → δ	10	20	30	40	50	60	70	80	90
10	72.25	65.93	62.62	60.50	58.94	57.69	56.60	55.59	54.59
20	78.05	71.45	67.16	64.03	61.51	59.33	57.32	55.36	53.31
30	80.81	74.61	69.99	66.28	63.07	60.11	57.24	54.28	51.04
40	82.35	76.52	71.76	67.63	63.82	60.13	56.34	52.25	47.57
50	83.28	77.70	72.80	68.27	63.87	59.37	54.54	49.08	42.55
60	83.85	78.38	73.28	68.30	63.22	57.76	51.64	44.43	35.53
70	84.16	78.67	73.27	67.72	61.78	55.12	47.33	37.86	26.03
80	84.28	78.61	72.76	66.45	59.37	51.10	41.12	28.84	13.93
90	84.22	78.21	71.68	64.30	55.64	45.19	32.40	17.09	.00
100	83.98	77.39	69.85	60.92	50.02	36.62	20.65	3.13	−13.93
110	83.52	76.01	66.94	55.66	41.55	24.58	6.15	−11.29	−26.03
120	82.74	73.77	62.26	47.34	28.94	8.96	−9.34	−24.12	−35.53
130	81.46	70.03	54.37	33.97	11.46	−8.36	−23.36	−34.31	−42.55
140	79.24	63.26	40.27	13.58	−8.74	−24.09	−34.46	−41.87	−47.57
150	74.90	49.19	15.27	−11.20	−26.81	−36.26	−42.60	−47.28	−51.04
160	63.88	16.50	−17.18	−32.27	−40.09	−44.94	−48.37	−51.04	−53.31
170	17.25	−29.88	−41.41	−46.28	−49.05	−50.93	−52.35	−53.53	−54.59
180	−55.00	−55.00	−55.00	−55.00	−55.00	−55.00	−55.00	−55.00	−55.00

(HA) \ (Az) → δ	100	110	120	130	140	150	160	170	180
10	53.53	52.35	50.93	49.05	46.28	41.41	29.88	−17.25	−90.00
20	51.04	48.37	44.94	40.09	32.27	17.18	−16.50	−63.88	−90.00
30	47.28	42.60	36.26	26.81	11.20	−15.27	−49.19	−74.90	−90.00
40	41.87	34.46	24.09	8.74	−13.58	−40.27	−63.26	−79.24	−90.00
50	34.31	23.36	8.36	−11.46	−33.97	−54.37	−70.03	−81.46	−90.00
60	24.12	9.34	−8.96	−28.94	−47.34	−62.26	−73.77	−82.74	−90.00
70	11.29	−6.15	−24.58	−41.55	−55.66	−66.94	−76.01	−83.52	−90.00
80	−3.13	−20.65	−36.62	−50.02	−60.92	−69.85	−77.39	−83.98	−90.00
90	−17.09	−32.40	−45.19	−55.64	−64.30	−71.68	−78.21	−84.22	−90.00
100	−28.84	−41.12	−51.10	−59.37	−66.45	−72.76	−78.61	−84.28	−90.00
110	−37.86	−47.33	−55.12	−61.78	−67.72	−73.27	−78.67	−84.16	−90.00
120	−44.43	−51.64	−57.76	−63.22	−68.30	−73.28	−78.38	−83.85	−90.00
130	−49.08	−54.54	−59.37	−63.87	−68.27	−72.80	−77.70	−83.28	−90.00
140	−52.25	−56.34	−60.13	−63.82	−67.63	−71.76	−76.52	−82.35	−90.00
150	−54.28	−57.24	−60.11	−63.07	−66.28	−69.99	−74.61	−80.81	−90.00
160	−55.36	−57.32	−59.33	−61.51	−64.03	−67.16	−71.45	−78.05	−90.00
170	−55.59	−56.60	−57.69	−58.94	−60.50	−62.62	−65.93	−72.25	−90.00
180	−55.00	−55.00	−55.00	−55.00	−55.00	−55.00	−55.00	−55.00	

5.6(57°) Table 5.5

Table of coordinates of points for plotting azimuth curves relating to latitude 57°N. The relation used is

$$\tan \delta = \frac{\sin \phi \cos(\text{HA}) + \sin(\text{HA}) \cot(\text{Az})}{\cos \phi}$$

δ (HA) \ (Az)	10	20	30	40	50	60	70	80	90
10	73.26	67.32	64.20	62.20	60.73	59.54	58.51	57.55	56.60
20	78.71	72.50	68.47	65.51	63.13	61.07	59.17	57.30	55.35
30	81.31	75.46	71.12	67.61	64.58	61.78	59.05	56.23	53.13
40	82.76	77.26	72.77	68.86	65.26	61.75	58.14	54.22	49.71
50	83.64	78.36	73.73	69.44	65.26	60.97	56.34	51.07	44.71
60	84.17	78.99	74.16	69.43	64.58	59.36	53.44	46.41	37.59
70	84.46	79.25	74.12	68.84	63.14	56.71	49.11	39.72	27.77
80	84.57	79.19	73.61	67.57	60.74	52.67	42.78	30.38	14.97
90	84.51	78.79	72.54	65.44	57.01	46.67	33.75	17.94	.00
100	84.28	77.99	70.76	62.09	51.34	37.83	21.34	2.94	−14.97
110	83.84	76.65	67.89	56.82	42.65	25.15	5.78	−12.54	−27.77
120	83.09	74.47	63.25	48.37	29.44	8.42	−10.82	−26.08	−37.59
130	81.85	70.82	55.34	34.47	10.78	−10.08	−25.54	−36.57	−44.71
140	79.72	64.14	40.85	12.78	−10.72	−26.48	−36.87	−44.17	−49.71
150	75.52	49.93	14.39	−13.47	−29.39	−38.78	−44.98	−49.52	−53.13
160	64.69	15.55	−19.76	−34.94	−42.62	−47.32	−50.62	−53.19	−55.35
170	16.26	−32.64	−43.96	−48.66	−51.32	−53.11	−54.47	−55.60	−56.60
180	−57.00	−57.00	−57.00	−57.00	−57.00	−57.00	−57.00	−57.00	−57.00

δ (HA) \ (Az)	100	110	120	130	140	150	160	170	180
10	55.60	54.47	53.11	51.32	48.66	43.96	32.64	−16.26	−90.00
20	53.19	50.62	47.32	42.62	34.94	19.76	−15.55	−64.69	−90.00
30	49.52	44.98	38.78	29.39	13.47	−14.39	−49.93	−75.52	−90.00
40	44.17	36.87	26.48	10.72	−12.78	−40.85	−64.14	−79.72	−90.00
50	36.57	25.54	10.08	−10.78	−34.47	−55.34	−70.82	−81.85	−90.00
60	26.08	10.82	−8.42	−29.44	−48.37	−63.25	−74.47	−83.09	−90.00
70	12.54	−5.78	−25.15	−42.65	−56.82	−67.89	−76.65	−83.84	−90.00
80	−2.94	−21.34	−37.83	−51.34	−62.09	−70.76	−77.99	−84.28	−90.00
90	−17.94	−33.75	−46.67	−57.01	−65.44	−72.54	−78.79	−84.51	−90.00
100	−30.38	−42.78	−52.67	−60.74	−67.57	−73.61	−79.19	−84.67	−90.00
110	−39.72	−49.11	−56.71	−63.14	−68.84	−74.12	−79.25	−84.46	−90.00
120	−46.41	−53.44	−59.36	−64.58	−69.43	−74.16	−78.99	−84.17	−90.00
130	−51.07	−56.34	−60.97	−65.26	−69.44	−73.73	−78.36	−83.64	−90.00
140	−54.22	−58.14	−61.75	−65.26	−68.86	−72.77	−77.26	−82.76	−90.00
150	−56.23	−59.05	−61.78	−64.58	−67.61	−71.12	−75.46	−81.31	−90.00
160	−57.30	−59.17	−61.07	−63.13	−65.51	−68.47	−72.50	−78.71	−90.00
170	−57.55	−58.51	−59.54	−60.73	−62.20	−64.20	−67.32	−73.26	−90.00
180	−57.00	−57.00	−57.00	−57.00	−57.00	−57.00	−57.00	−57.00	

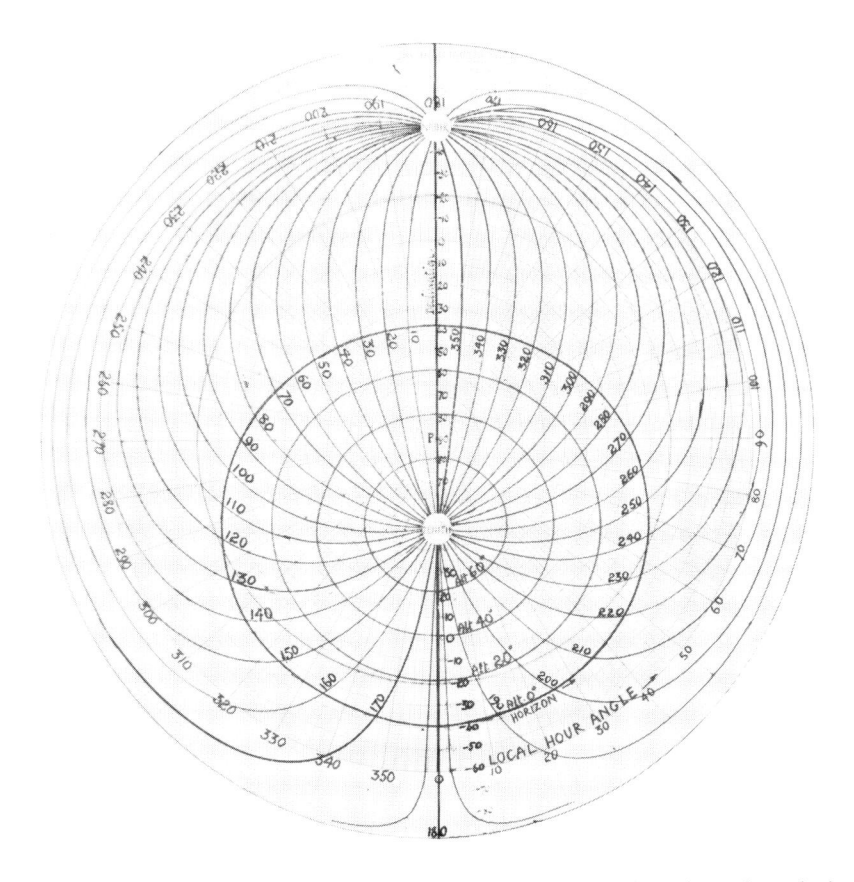

Fig. 5.1. The azimuth curves based on the relation
$$\tan \delta = \frac{\sin \phi \cos(\text{HA}) + \sin(\text{HA}) \cot(\text{Az})}{\cos \phi}$$
are each continuous loops from zenith. South to nadir and north back to zenith. They do not stop at the horizon.

5.7 The Tables 5.7–5.11 (40°, 51°, 53°, 55° and 57°), were drawn up by calculating the HA's in the expression above, for a range of declinations from +80° to −30 while keeping the *altitude* fixed for a particular altitude curve.

It will be noticed that only a limited range of declinations and altitudes are admissable, as the calculator rightly rejects HA's combined with declinations that produce HA's with cosines >1.

It is a matter for special interest to draw the curve for zero altitude, i.e., for the horizon, as it has approximate elliptical shape and not circular as might perhaps be expected.

5.8 The Horizon Curve

The coordinates for the altitude curve (alt) = 0, i.e., for the horizon are given by the altitude relation [3.7(1)],

$$\cos(\mathrm{HA}) = \frac{\sin(\mathrm{alt}) - \sin\phi\sin\delta}{\cos\phi\cos\delta} \quad \text{but with} \quad \sin(\mathrm{alt}) = 0,$$

then $\quad \cos(\mathrm{HA}) = -\tan\phi\tan\delta.$

The calculator gives the following table from $\tan\delta = -\cos(\mathrm{HA})/\tan\phi$ with $\phi = 51$ which makes the horizon curve easy to draw (see Section 3.14).

Table 5.6

(HA)		δ	
0	and 360	−39.00	
10	and 350	−38.57	
20	and 340	−37.27	
30	and 330	−35.04	
40	and 320	−31.81	
50	and 310	−27.59	
57.62	and 302.38	−23.44	Winter solstice
60	and 300	−22.04	
70	and 290	−15.48	
80	and 280	− 8.00	
90	and 270	0.00	
100	and 260	8.00	
110	and 250	15.48	
120	and 240	22.04	
122.37	and 237.62	23.44	Summer solstice
130	and 230	27.59	
140	and 220	31.81	
150	and 210	35.04	
160	and 200	37.27	
170	and 190	38.51	
180		39.00	

From this curve the times and azimuths of the risings and settings of all bodies having declinations between + and −39° can be found. This horizon curve is elliptical in shape on account of the distortion caused by representing the celestial sphere on a plane surface polar diagram. The HA for the Sun when corrected for longitude and the equation of time will give the local time of sunrise and sunset. A polar type planisphere has to have the centre of its horizon ellipse (the zenith) at a distance of $(90-\phi)$ from the poles where ϕ is the latitude of the place.

Altitude and Azimuth Lines

5.9(40°)

Table 5.7
Table of coordinates of points for plotting altitude curves or almucantars for latitude 40°N (see Section 5.7). The relation used is

$$\cos(HA) = \frac{\sin(alt) - \sin\phi \sin\delta}{\cos\phi \cos\delta}$$

(HA)\(Dec)	(Alt)→ 5	10	15	20	25	30	35	40	45	50	55	60	65	70	75	80	85
80	.00	.00	.00	.00	.00	179.98	116.54	85.79	56.16	.02	.00	.00	.00	.00	.00	.00	.00
75	.00	.00	.00	.00	179.99	127.57	103.80	83.66	64.22	42.93	.02	.00	.00	.00	.00	.00	.00
70	.00	.00	.00	179.99	133.82	113.39	96.67	81.49	66.83	51.80	34.81	.02	.00	.00	.00	.00	.00
65	.00	.00	179.99	137.99	119.61	104.78	91.59	79.28	67.38	55.48	43.05	28.89	.01	.00	.00	.00	.00
60	.00	179.99	141.04	124.08	103.68	98.51	87.47	77.01	66.87	56.86	46.74	36.13	24.10	.00	.00	.00	.00
55	179.99	143.43	127.54	103.68	103.68	93.46	83.85	74.66	65.74	56.97	48.24	39.41	30.20	19.90	.00	.00	.00
50	145.39	130.34	118.32	107.78	98.15	89.12	80.51	72.22	64.15	56.24	48.43	40.64	32.80	24.72	15.92	.00	.00
45	132.70	121.23	111.18	101.99	93.38	85.18	77.30	69.66	62.21	54.89	47.69	40.56	33.48	26.40	19.24	11.77	.01
40	123.75	114.09	105.25	96.96	89.08	81.49	74.14	66.97	59.94	53.04	46.23	39.49	32.82	26.20	19.62	13.07	6.53
35	116.66	108.11	100.08	92.44	85.07	77.92	70.94	64.10	57.36	50.71	44.12	37.57	31.05	24.50	17.87	10.93	.01
30	110.68	102.87	95.41	88.22	81.22	74.38	67.66	61.02	54.45	47.91	41.38	34.82	28.16	21.25	13.70	.00	.00
25	105.41	98.12	91.06	84.18	77.44	70.80	64.22	57.69	51.16	44.59	37.95	31.12	23.92	15.80	.00	.00	.00
20	100.62	93.68	86.90	80.23	73.64	67.10	60.57	54.02	47.40	40.64	33.64	26.15	17.52	.00	.00	.00	.00
15	96.15	89.44	82.82	76.27	69.74	63.20	56.61	49.92	43.05	35.86	28.09	18.99	.01	.00	.00	.00	.00
10	91.86	85.28	78.75	72.22	65.65	59.01	52.24	45.24	37.88	29.83	20.30	0.1	.00	.00	.00	.00	.00
5	87.66	81.13	74.59	67.99	61.29	54.42	47.30	39.75	31.44	21.50	.00	.00	.00	.00	.00	.00	.00
0	83.47	76.90	70.25	63.48	56.52	49.25	41.52	32.95	22.62	.01	.00	.00	.00	.00	.00	.00	.00
−5	79.19	72.48	65.63	58.56	51.16	43.23	34.41	23.69	.00	.00	.00	.00	.00	.00	.00	.00	.00
−10	74.72	67.78	60.59	53.04	44.91	35.83	24.73	.00	.00	.00	.00	.00	.00	.00	.00	.00	.00
−15	69.96	62.64	54.93	46.60	37.25	25.77	.01	.00	.00	.00	.00	.00	.00	.00	.00	.00	.00
−20	64.76	56.86	48.32	38.69	26.81	.01	.00	.00	.00	.00	.00	.00	.00	.00	.00	.00	.00
−25	58.88	50.10	40.18	27.88	.00	.00	.00	.00	.00	.00	.00	.00	.00	.00	.00	.00	.00
−30	51.99	41.74	29.00	.00	.00	.00	.00	.00	.00	.00	.00	.00	.00	.00	.00	.00	.00
−35	43.41	30.20	.00	.00	.00	.00	.00	.00	.00	.00	.00	.00	.00	.00	.00	.00	.00
−40	31.50	.00	.00	.00	.00	.00	.00	.00	.00	.00	.00	.00	.00	.00	.00	.00	.00

Positional Astronomy and Astro-Navigation Made Easy

5.9(51°)

Table 5.8
Table of coordinates of points for plotting *altitude* curves or almucantars for use on star charts of various types, Polar, Cartesian, Mercator or Stereographic projection. The relation used is

$$\cos(\mathrm{HA}) = \frac{\sin(\mathrm{alt}) - \sin\phi\sin\delta}{\cos\phi\cos\delta} \quad \text{where } \phi = 51° \text{ N or S}$$

(HA)(Dec)\\(Alt)	5	10	15	20	25	30	35	40	45	50	55	60	65	70	75	80	85
80	.00	.00	.00	.00	.00	.00	.00	.00	.00	89.63	60.50	22.87	.00	.00	.00	.00	.00
75	.00	.00	.00	.00	.00	.00	.00	131.48	122.20	84.58	65.14	44.91	17.14	.00	.00	.00	.00
70	.00	.00	.00	.00	.00	.00	136.72	113.98	105.51	80.43	65.61	50.90	35.13	13.36	.00	.00	.00
65	.00	.00	.00	.00	.00	140.20	119.45	103.38	96.18	76.58	64.42	52.56	40.59	27.76	10.40	.00	.00
60	.00	.00	.00	.00	142.73	123.36	108.42	95.51	89.40	72.81	62.33	52.17	42.15	32.06	21.43	7.76	.00
55	.00	.00	.00	144.70	126.36	112.24	100.06	89.02	83.78	68.99	59.62	50.54	41.65	32.89	24.17	15.28	4.99
50	.00	.00	146.29	128.77	115.27	103.63	93.08	83.26	78.74	65.04	56.41	48.00	39.76	31.65	23.63	15.67	7.71
45	.00	147.64	130.79	117.79	106.57	96.39	86.90	77.90	73.96	60.89	52.71	44.66	36.70	28.74	20.65	11.99	.00
40	148.81	132.53	119.96	109.07	99.18	89.95	81.17	72.71	69.26	56.44	48.47	40.52	32.46	24.07	14.66	.00	.00
35	134.08	121.86	111.26	101.61	92.57	83.96	75.64	67.53	64.50	51.59	43.59	35.39	26.70	16.63	.00	.00	.00
30	123.58	113.23	103.77	94.90	86.42	78.20	70.16	62.20	59.54	46.16	37.81	28.83	18.20	.00	.00	.00	.00
25	115.03	105.75	97.01	88.64	80.50	72.49	64.55	56.55	54.24	39.89	30.64	19.52	.00	.00	.00	.00	.00
20	107.58	98.96	90.68	82.59	74.62	66.67	58.64	50.40	48.40	32.23	20.66	.00	.00	.00	.00	.00	.00
15	100.81	92.59	84.56	76.60	68.63	60.55	52.22	43.40	41.73	21.67	.00	.00	.00	.00	.00	.00	.00
10	94.42	86.42	78.47	70.48	62.34	53.91	44.95	34.97	33.66	.00	.00	.00	.00	.00	.00	.00	.00
5	88.22	80.27	72.25	64.05	55.52	46.41	36.21	23.47	22.60	.00	.00	.00	.00	.00	.00	.00	.00
0	82.04	73.98	65.72	57.08	47.81	37.39	24.30	.00	.00	.00	.00	.00	.00	.00	.00	.00	.00
−5	75.70	67.35	58.61	49.19	38.54	25.10	.00	.00	.00	.00	.00	.00	.00	.00	.00	.00	.00
−10	69.00	60.14	50.55	39.68	25.89	.00	.00	.00	.00	.00	.00	.00	.00	.00	.00	.00	.00
−15	61.69	51.94	40.83	26.68	.00	.00	.00	.00	.00	.00	.00	.00	.00	.00	.00	.00	.00
−20	53.36	42.00	27.49	.00	.00	.00	.00	.00	.00	.00	.00	.00	.00	.00	.00	.00	.00
−25	43.23	28.32	.00	.00	.00	.00	.00	.00	.00	.00	.00	.00	.00	.00	.00	.00	.00
−30	29.20	.00	.00	.00	.00	.00	.00	.00	.00	.00	.00	.00	.00	.00	.00	.00	.00
−35	.00	.00	.00	.00	.00	.00	.00	.00	.00	.00	.00	.00	.00	.00	.00	.00	.00
−40	.00	.00	.00	.00	.00	.00	.00	.00	.00	.00	.00	.00	.00	.00	.00	.00	.00

Altitude and Azimuth Lines

5.9(53°)

Table 5.9
Table of coordinates of points for plotting altitude curves or almucantars for latitude 53°N or S.

(HA) (Dec)	(Alt) 5	10	15	20	25	30	35	40	45	50	55	60	65	70	75	80	85
80	.00	.00	.00	.00	.00	.00	.00	.00	139.44	101.29	71.79	40.45	.00	.00	.00	.00	.00
75	.00	.00	.00	.00	.00	.00	.00	145.67	114.39	91.98	72.16	52.60	30.01	.00	.00	.00	.00
70	.00	.00	.00	.00	.00	.00	149.25	121.54	102.16	85.66	70.51	55.85	40.79	23.18	.00	.00	.00
65	.00	.00	.00	.00	.00	151.64	126.21	108.58	93.77	80.44	67.98	56.00	44.15	31.92	17.83	.00	.00
60	.00	.00	.00	.00	153.38	129.56	113.10	99.34	87.05	75.68	64.93	54.58	44.49	34.48	24.28	13.02	.00
55	.00	.00	.00	154.74	132.14	116.53	103.51	91.90	81.18	71.10	61.45	52.15	43.09	34.20	25.44	16.71	7.80
50	.00	.00	155.85	134.22	119.28	106.80	95.67	85.40	75.74	66.50	57.59	48.91	40.42	32.04	23.73	15.36	6.43
45	.00	156.78	135.96	121.56	109.51	98.75	88.81	79.43	70.45	61.76	53.28	44.92	36.61	28.22	19.47	9.19	.00
40	157.59	137.46	123.51	111.82	101.35	91.66	82.49	73.69	65.15	56.76	48.45	40.09	31.53	22.36	10.98	.00	.00
35	138.80	125.24	113.84	103.62	94.12	85.12	76.45	68.00	59.66	51.34	42.91	34.16	24.60	12.33	.00	.00	.00
30	126.79	115.66	105.64	96.31	87.44	78.86	70.47	62.15	53.80	45.28	36.34	26.43	13.40	.00	.00	.00	.00
25	117.32	107.48	98.30	89.53	81.02	72.67	64.36	55.97	47.34	38.22	27.99	14.31	.00	.00	.00	.00	.00
20	109.20	110.13	91.45	83.00	74.67	66.35	57.91	49.18	39.88	29.36	15.10	.00	.00	.00	.00	.00	.00
15	101.87	93.26	84.86	76.54	68.20	59.70	50.87	41.39	30.59	15.80	.00	.00	.00	.00	.00	.00	.00
10	94.99	86.62	78.30	69.93	61.37	52.44	42.80	31.73	16.45	.00	.00	.00	.00	.00	.00	.00	.00
5	88.32	80.01	71.60	62.97	53.93	44.12	32.79	17.05	.00	.00	.00	.00	.00	.00	.00	.00	.00
0	81.67	73.23	64.53	55.37	45.39	33.82	17.62	.00	.00	.00	.00	.00	.00	.00	.00	.00	.00
−5	74.84	66.06	56.78	46.64	34.81	18.18	.00	.00	.00	.00	.00	.00	.00	.00	.00	.00	.00
−10	67.60	58.20	47.88	35.80	18.73	.00	.00	.00	.00	.00	.00	.00	.00	.00	.00	.00	.00
−15	59.63	49.13	36.79	19.28	.00	.00	.00	.00	.00	.00	.00	.00	.00	.00	.00	.00	.00
−20	50.42	37.81	19.84	.00	.00	.00	.00	.00	.00	.00	.00	.00	.00	.00	.00	.00	.00
−25	38.87	20.42	.00	.00	.00	.00	.00	.00	.00	.00	.00	.00	.00	.00	.00	.00	.00
−30	21.03	.00	.00	.00	.00	.00	.00	.00	.00	.00	.00	.00	.00	.00	.00	.00	.00
−35	.00	.00	.00	.00	.00	.00	.00	.00	.00	.00	.00	.00	.00	.00	.00	.00	.00
−40	.00	.00	.00	.00	.00	.00	.00	.00	.00	.00	.00	.00	.00	.00	.00	.00	.00

5.9 (55°)

Table 5.10
Table of coordinates of points for plotting altitude curves or almucantars for latitude 55°N or S.

(HA)\(Alt)(Dec)	5	10	15	20	25	30	35	40	45	50	55	60	65	70	75	80	85
80	.00	.00	.00	.00	.00	.00	.00	.00	.00	.00	.00	.00	.00	.00	.00	.00	.00
75	.00	.00	.00	.00	.00	.00	.00	.00	.00	.00	.00	.00	.00	.00	.00	.00	.00
70	.00	.00	.00	.00	.00	.00	.00	.00	.00	.00	.00	.00	.00	29.97	.00	.00	.00
65	.00	.00	.00	.00	.00	179.98	179.98	179.98	179.98	114.10	82.82	53.45	.00	.00	.00	.00	.00
60	.00	.00	.00	.00	179.98	136.90	134.14	130.33	124.52	99.77	79.16	59.75	39.18	.00	.00	.00	.00
55	.00	.00	.00	179.98	139.03	121.32	118.27	114.26	108.62	91.08	75.41	60.61	45.89	35.52	22.76	.00	.00
50	.00	.00	.00	140.74	123.76	110.23	107.23	103.43	98.37	84.40	71.54	59.34	47.46	36.58	26.56	16.20	.01
45	.00	.00	179.98	125.79	112.71	101.26	98.41	94.92	90.46	78.61	67.50	56.90	46.64	35.25	26.31	17.48	8.72
40	179.99	179.98	142.19	114.83	103.68	90.80	90.80	87.62	83.70	73.21	63.24	53.65	44.34	32.14	23.38	14.28	.01
35	144.54	143.43	127.54	105.79	95.77	86.32	83.85	80.98	77.53	67.93	58.68	49.69	40.87	27.28	17.55	.00	.00
30	130.47	129.08	116.69	97.82	88.50	79.51	77.25	74.66	71.62	62.57	53.73	45.00	36.25	19.90	.00	.00	.00
25	119.89	118.36	107.67	90.46	81.55	72.79	70.72	68.40	65.74	56.97	48.24	39.41	30.20	.00	.00	.00	.00
20	110.98	109.38	99.68	83.41	74.67	65.93	64.06	62.00	59.67	50.92	41.97	32.52	21.73	.00	.00	.00	.00
15	103.02	101.40	92.27	76.43	67.66	58.68	57.02	55.21	53.20	44.14	34.46	23.23	.01	.00	.00	.00	.00
10	95.60	93.97	85.15	69.29	60.24	50.70	49.26	47.72	46.03	36.13	24.52	.01	.00	.00	.00	.00	.00
5	88.42	86.81	78.09	61.73	52.07	41.40	40.22	38.97	37.62	25.65	.01	.00	.00	.00	.00	.00	.00
0	81.26	79.69	70.85	53.40	42.54	29.34	28.49	27.61	26.67	.01	.00	.00	.00	.00	.00	.00	.00
−5	73.89	72.38	63.18	43.65	30.17	.01	.01	.01	.01	.00	.00	.00	.00	.00	.00	.00	.00
−10	66.04	64.61	54.70	30.98	.01	.00	.00	.00	.00	.00	.00	.00	.00	.00	.00	.00	.00
−15	57.32	56.00	44.76	.01	.00	.00	.00	.00	.00	.00	.00	.00	.00	.00	.00	.00	.00
−20	47.04	45.89	31.81	.00	.00	.00	.00	.00	.00	.00	.00	.00	.00	.00	.00	.00	.00
−25	33.53	32.65	.01	.00	.00	.00	.00	.00	.00	.00	.00	.00	.00	.00	.00	.00	.00
−30	.01	.01	.00	.00	.00	.00	.00	.00	.00	.00	.00	.00	.00	.00	.00	.00	.00
−35	.00	.00	.00	.00	.00	.00	.00	.00	.00	.00	.00	.00	.00	.00	.00	.00	.00
−40	.00	.00	.00	.00	.00	.00	.00	.00	.00	.00	.00	.00	.00	.00	.00	.00	.00

5.9 (57°)

Altitude and Azimuth Lines

Table 5.11
Table of coordinates of points for plotting altitude curves or almucantars for latitude 57°N or S.

(HA)(Dec) \ (Alt)	5	10	15	20	25	30	35	40	45	50	55	60	65	70	75	80	85
80	.00	.00	.00	.00	.00	.00	.00	.00	.00	129.29	94.11	64.92	31.80	.00	.00	.00	.00
75	.00	.00	.00	.00	.00	.00	.00	.00	136.94	108.21	86.32	66.62	46.96	23.16	.00	.00	.00
70	.00	.00	.00	.00	.00	.00	.00	141.26	115.77	96.80	80.40	65.27	50.61	35.53	17.32	.00	.00
65	.00	.00	.00	.00	.00	.00	144.13	120.64	103.31	88.52	75.13	62.60	50.56	38.71	26.59	12.50	.00
60	.00	.00	.00	.00	.00	146.21	124.12	107.86	94.04	81.61	70.07	59.13	48.63	38.41	28.37	18.33	7.67
55	.00	.00	.00	.00	147.81	126.77	111.29	98.14	86.31	75.34	64.97	55.03	45.41	36.01	26.76	17.58	8.20
50	.00	.00	.00	149.11	128.90	114.01	101.35	89.95	79.36	69.33	59.69	50.31	41.09	31.89	22.49	12.07	.00
45	.00	.00	150.21	130.68	116.26	103.98	92.90	82.58	72.77	63.30	54.04	44.86	35.56	25.82	14.47	.00	.00
40	.00	151.15	132.20	118.19	106.21	95.38	85.26	75.61	66.25	57.05	47.83	38.41	28.34	16.22	.00	.00	.00
35	151.99	133.55	119.87	108.16	97.52	87.56	78.03	68.74	59.56	50.30	40.72	30.35	17.60	.00	.00	.00	.00
30	134.77	121.39	109.90	99.43	89.60	80.15	70.91	61.72	52.40	42.69	32.04	18.73	.00	.00	.00	.00	.00
25	122.78	111.49	101.17	91.44	82.06	72.85	63.64	54.26	44.40	33.50	19.70	.00	.00	.00	.00	.00	.00
20	112.97	102.78	93.14	83.81	74.62	65.39	55.93	45.93	34.80	20.56	.00	.00	.00	.00	.00	.00	.00
15	104.30	94.73	85.45	76.26	67.00	57.46	47.34	35.98	21.30	.00	.00	.00	.00	.00	.00	.00	.00
10	96.26	87.01	77.82	68.52	58.91	48.65	37.07	22.04	.00	.00	.00	.00	.00	.00	.00	.00	.00
5	88.51	79.32	69.98	60.29	49.89	38.11	22.72	.00	.00	.00	.00	.00	.00	.00	.00	.00	.00
0	80.79	71.41	61.63	51.10	39.11	23.36	.00	.00	.00	.00	.00	.00	.00	.00	.00	.00	.00
−5	72.82	62.95	52.28	40.08	23.99	.00	.00	.00	.00	.00	.00	.00	.00	.00	.00	.00	.00
−10	64.28	53.47	41.06	24.61	.00	.00	.00	.00	.00	.00	.00	.00	.00	.00	.00	.00	.00
−15	54.67	42.04	25.23	.00	.00	.00	.00	.00	.00	.00	.00	.00	.00	.00	.00	.00	.00
−20	43.05	25.87	.00	.00	.00	.00	.00	.00	.00	.00	.00	.00	.00	.00	.00	.00	.00
−25	26.54	.00	.00	.00	.00	.00	.00	.00	.00	.00	.00	.00	.00	.00	.00	.00	.00
−30	.00	.00	.00	.00	.00	.00	.00	.00	.00	.00	.00	.00	.00	.00	.00	.00	.00
−35	.00	.00	.00	.00	.00	.00	.00	.00	.00	.00	.00	.00	.00	.00	.00	.00	.00
−40	.00	.00	.00	.00	.00	.00	.00	.00	.00	.00	.00	.00	.00	.00	.00	.00	.00

Positional Astronomy and Astro-Navigation Made Easy

Fig. 5.1 shows the azimuth curves and altitude curves constructed on polar graph paper of the same declination scale as the star chart or planisphere. The curves are then reproduced on a transparency such as that shown in the drawing and photographs.

Fig. 5.2 shows a drawing on polar graph paper of alt–azimuth curves within the horizon.

Fig. 5.5 shows the overlay in position over a planisphere which is rotated by a sidereal clock so that the *position*, *altitude* and *azimuth* of all stars can be depicted even if they are not visible in the night sky. The times of rising and setting can also be shown.

The horizon curve is a slightly distorted ellipse. It is important as only stars within its boundary are visible, provided the Sun is outside the boundary. The times of rising and setting can be found at a glance.

Fig. 5.6 An instructive project is to make a large scale class demonstration model of a rotating planisphere covered with an alt–azimuth overlay showing the positions of the stars of the northern or southern hemisphere, as shown in Fig. 5.6. For this model a large basic star chart (52 cm diameter) such as that published by the *Daily Telegraph* (London) is cut out and mounted on a hardboard circular disc which is free to rotate about its pole, the pivot being a small bolt secured to a base board, B, about 70 cm square.

Fixed graduated circles are drawn on the base plate, B, to show UT, the GHA and the SHA of the celestial bodies. The overlay of alt–azimuth curves is drawn on a circular disc of transparent acetate (56 cm in diameter) using 'permanent' felt pen ink. A convenient way of making overlays is to plot carefully one set of curves on good polar graph paper about 30 cm diameter, this is then photographed and the negative (35 mm) projected on to the circular acetate sheet, taking care that the magnification exactly fits the dimensions of the star chart. In the Fig. 5.6 the horizon curve has been emphasised by a thick dark line and stars

Altitude and Azimuth Lines

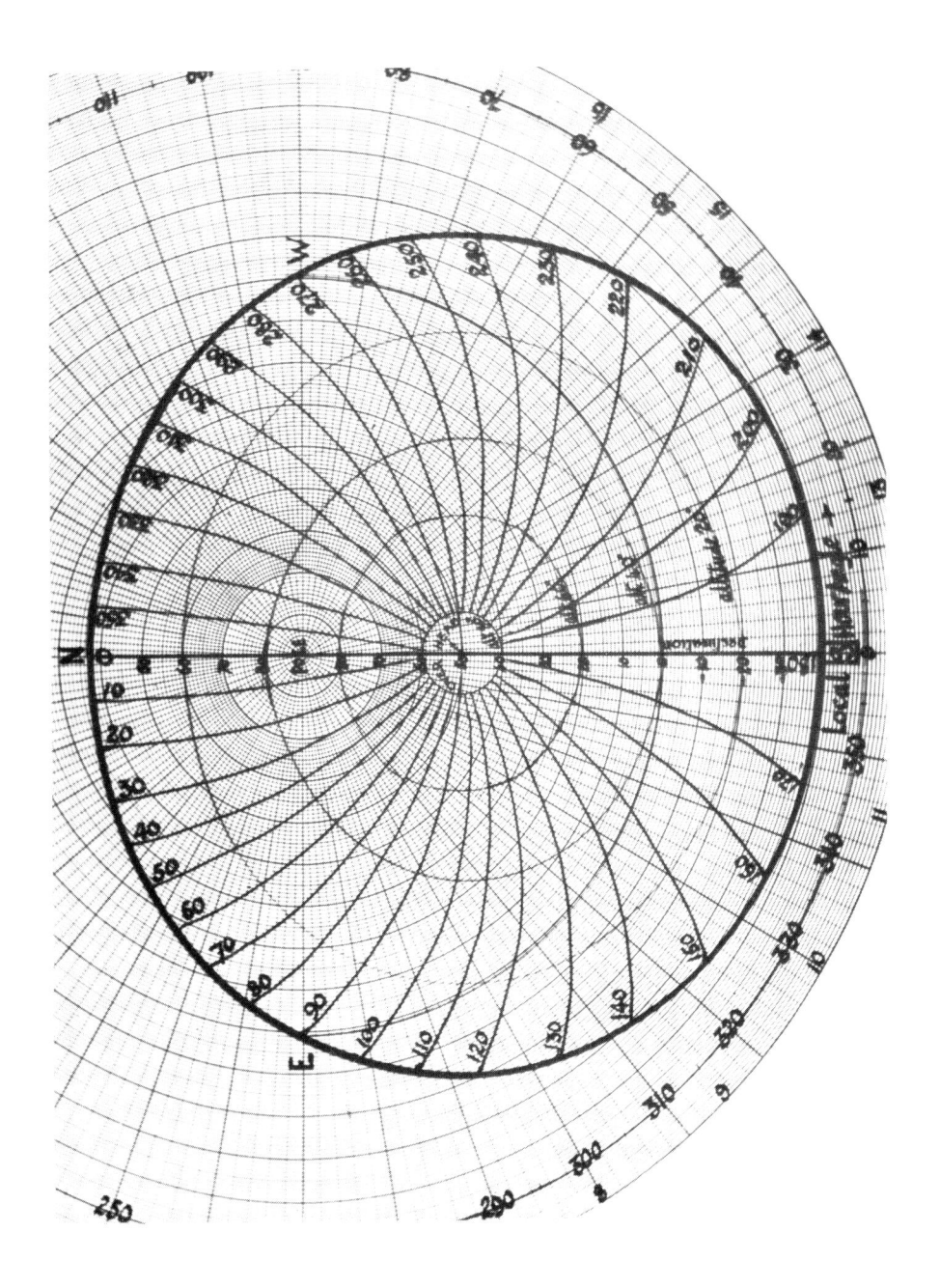

Fig. 5.2

within this horizon line are visible during the hours of darkness for an observer in latitude 51°.

An additional circular disc of acetate sheet can be superimposed on the star chart and attached to it to show the positions of the Sun, Moon, planets or comets on the date of demonstration. The positions can be marked in RA and declination by small coloured stars that adhere to the acetate sheet but can be easily moved from day to day thus serving the purpose of a planetarium.

The circular star chart can, as in Fig. 5.5, be coupled to a sidereal clock so that the alt–azimuth planisphere can show the positions in alt–azimuth and hour angle of all bodies in the visible sky at any time. Failing a sidereal clock coupling, the star disc can be rotated by hand about 15° every hour, or, as we have seen, the local mean time is put opposite the date as described in Section 2.4.

The series of tables under Section 5.6 connecting azimuths, declinations and hour angles for each latitude, and the series in Section 5.9 connecting hour angles, declinations and altitudes for each latitude, are applicable to both the northern and the southern hemispheres (see Section 3.7) because the azimuth relation expressed in [3.7(7)] is

$$\sin(\text{Az}) = \frac{\sin(\text{HA}) \cos \delta}{\cos(\text{alt})}$$

and this shows that the azimuth is independent of the latitude, and also that the sign of the declination does not affect the result since the cosine of δ is the same whether δ is positive or negative.

The altitude of a body is from [3.7(1)]

$$\sin(\text{alt}) = \sin \phi \sin \delta + \cos \phi \cos \delta \cos(\text{HA})$$

and it follows that if both ϕ and δ are south, i.e. *negative* then the altitude is the same as if both were positive.

The cosine of a negative angle is the same as that for a positive angle.

Further the relation for a negative ϕ and a positive δ yields the same result as for a positive ϕ and a negative δ.

The formulae and the calculator, when properly used, take good care of the signs and of both hemispheres.

Altitude and Azimuth Lines

Fig. 5.3. Planisphere with transparent overlay showing azimuths and altitudes.

Fig. 5.4 Shows the curve on tracing paper.

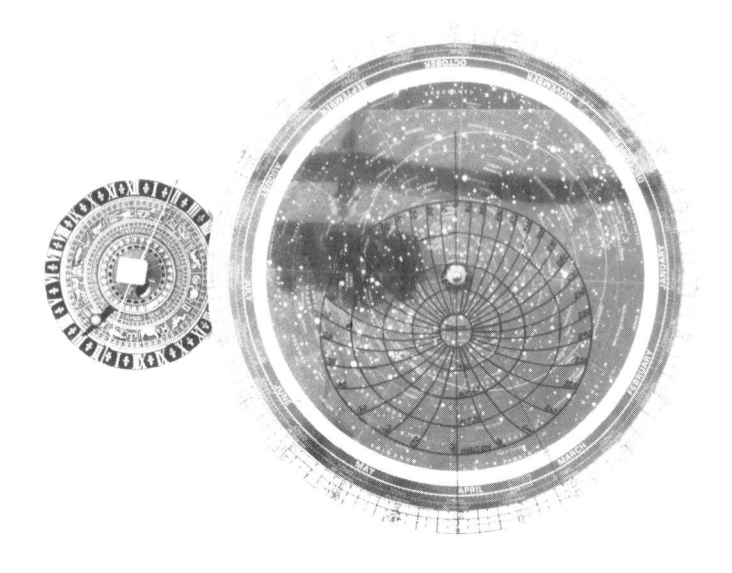

Positional Astronomy and Astro-Navigation Made Easy

Fig. 5.5. Planisphere with azimuth curve overlay—driven by a sidereal clock.

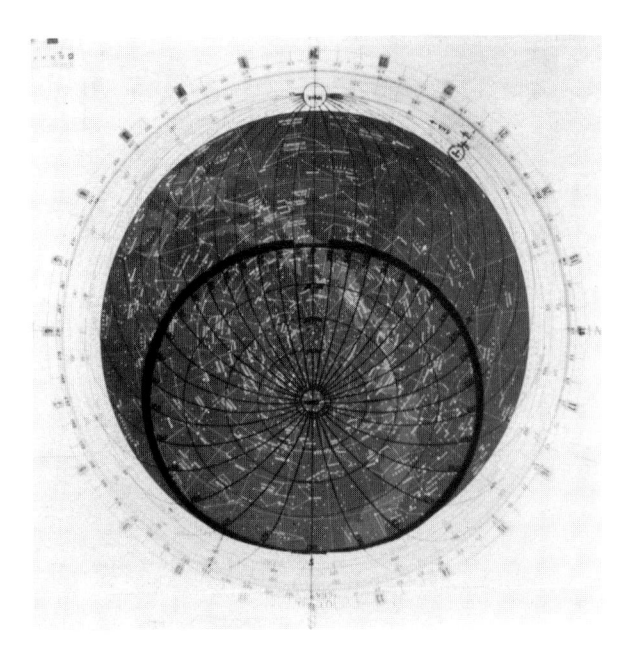

Fig. 5.6

5.10 Plotting the Values in the Tables on Centimetric Graph Paper using Rectangular Coordinates

The tabulated results can be represented very satisfactorily using ordinary rectangular graph paper. The result of doing this is shown in Fig. 5.7. The usefulness of this set of curves on centimetric graph paper is that it can be used to find the altitude and azimuth of any celestial body knowing its δ and its local hour angle. It is easy to use and does not require a star map, as the position on the graph paper can easily be marked by a fine pencil point, knowing the declination and the LHA. The graphs can be used, with a little interpolation, to find the local sun time or HA of the Sun, given the Sun's declination—or the date and the Sun's altitude.

For example, given the Sun's declination $+10°$ (28th August) and the Sun's altitude $20°$ (by observation) then the Sun's LHA is shown by the point P as $70°$ or $290°$, i.e. $4^h\ 40^m$ past noon or $4^h\ 40^m$ before noon. Therefore Sun's time $= 16^h\ 40^m$ or $07^h\ 20^m$. The graphs thus can serve as a sundial. See Section 7.11.

Two sets of alt–azimuth curves are shown in Fig. 5.7 for latitude 51°N (London) and Fig. 5.8 for latitude 40°N suitable for New York.

The coordinates of a celestial body, i.e. its LHA and declination, can of course be plotted on the polar type diagram shown in Fig. 5.2. The body's altitude and azimuth can then be found from these curves by a little interpolation. The polar diagram, however, presents a rather confusing pattern with the declination lines radiating from the pole and the azimuth lines radiating from the zenith. The simple cartesian coordinates with declinations along the Y axis and the LHA along the X axis will make the finding of altitudes and azimuths much easier. The LHA of a body can be found by methods described in Section 2.10, or by the method described above

$$\text{LHA} = \text{local sidereal time} - \text{RA*}$$

As an instructive alternative the LHA of a celestial body at any particular time for use on the cartesian diagrams can be found with fair accuracy from the Philips' planisphere. First set the planisphere with the date on the local time. This operation shows the LHA's of all the stars visible. Now use an ordinary plastic transparent 360° circular protractor to measure in degrees the hour angle of a particular star. The centre of the protractor should be at the pole. This measurement must be made from the south point of the meridian (or the 12 hour mark) in an anti-clockwise direction, E→W for an observer using a northern hemisphere planisphere but is measured in a clockwise direction E→W for an observer using a planisphere designed for southern latitudes.

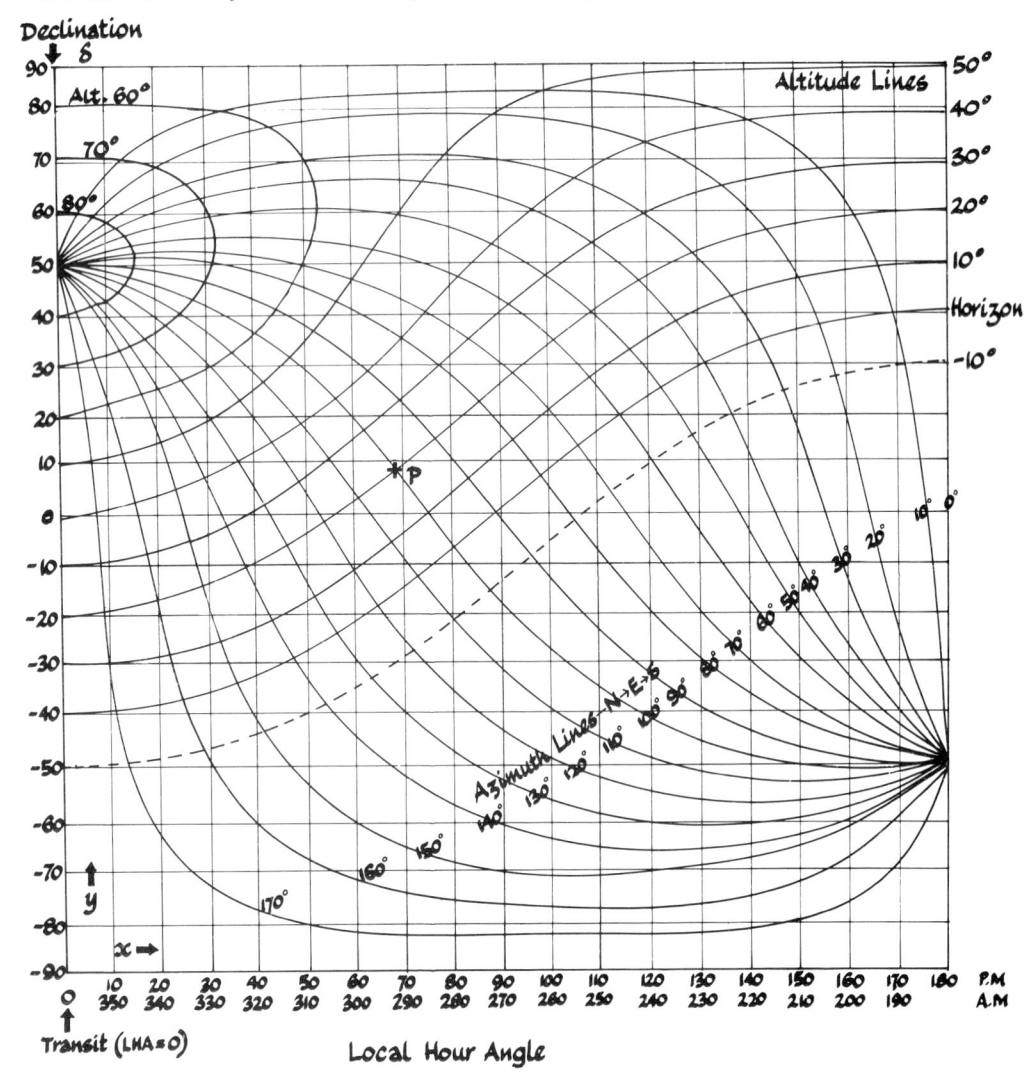

Positional Astronomy and Astro-Navigation Made Easy

Fig. 5.7. Altitude–azimuth curves for latitude 51°N drawn by plotting the coordinates in Tables 5.2 and 5.8*

To use this set of graphs to find where to look for a particular star, knowing its Right Ascension and declination and the date—(see Appendices IV and V).
(1) Find local sidereal time from the simple approximate relation
LST = LMT+(number of days since 21st September ×4 mins).
(2) Find the star's LHA from   LHA = LST−RA.
(3) From the graphs mark the point corresponding to x = LHA, y − declination.
(4) This point indicates with a little interpolation between the nearest curves, the altitude and azimuth of the star. For example, the point P shows altitude 20°, azimuth 100° for the Sun when the Sun's declination is +9 and LHA is 69°.

*Relations used for azimuth curves as in Sections 5.1 and 5.5.

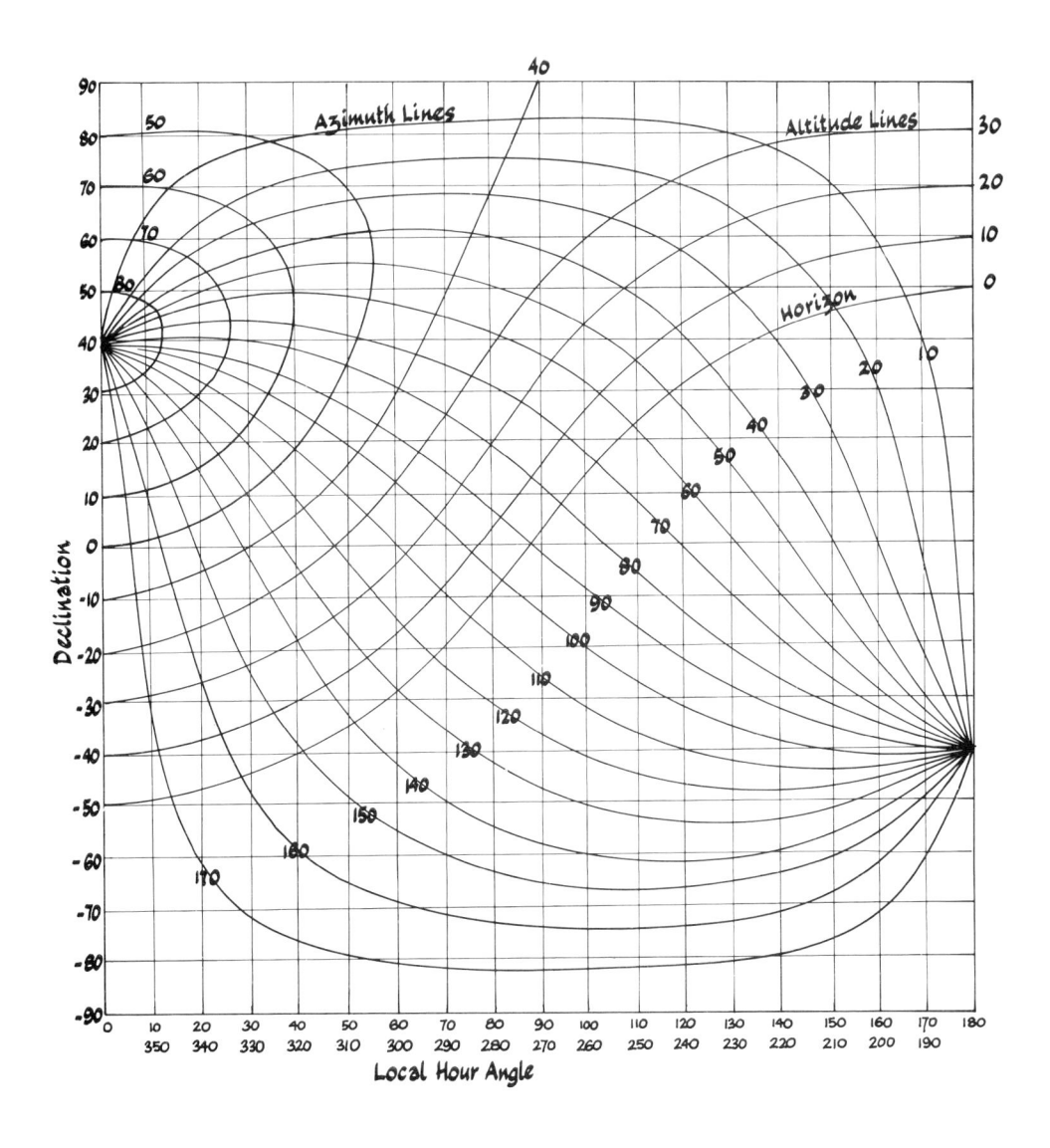

Fig. 5.8 These curves drawn for the latitude 40° north are equally applicable to the latitude 40° south provided hour angles are measured from the N–S meridian towards the west.

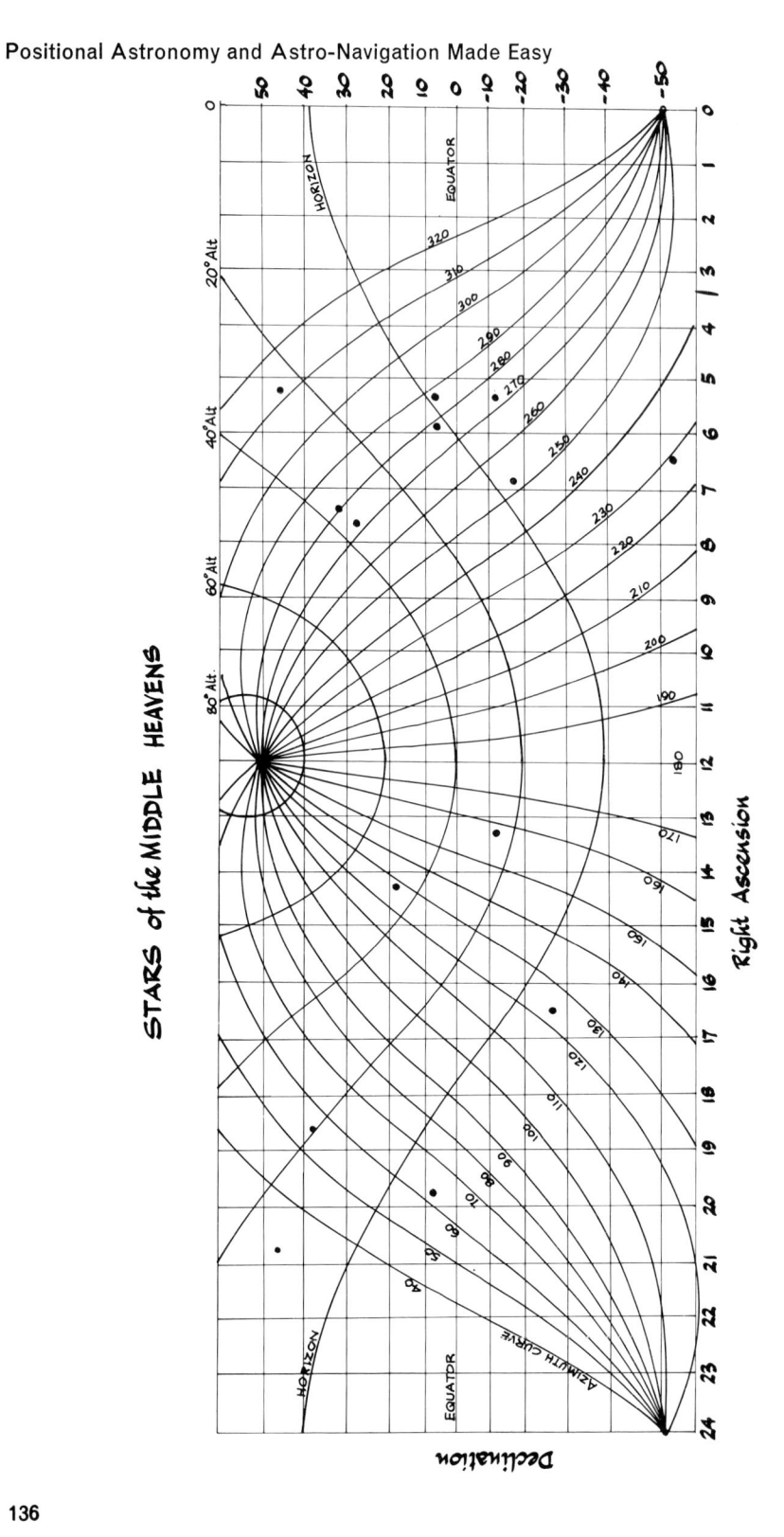

Positional Astronomy and Astro-Navigation Made Easy

Fig. 5.9. The Philips' *Chart of the Stars of the Middle Heavens* with an appropriate overlay transparency to give the altitudes and azimuths of stars or planets.
The values given in tables are used but making use of the Mercator's reference frame on which the δ's are shown as parallel straight lines, all perpendicular to the RA lines, but spaced by the same mathematical rules that apply to spacing of latitude lines on a Mercator nautical chart.

This is a simple way of finding the approximate LHA in degrees for use on the curves in Fig. 5.7 The edge of a sheet of paper will help to line up the pole, the star and the protractor reading. It is important to use the planisphere and the alt–azimuth diagram appropriate to the observer's latitude. A difference of 5° between the observer's latitude and the latitude for which the planisphere or alt–azimuth diagram is designed, can be accepted for approximate information.

This graticule is easy to use and can be accurately inscribed on a single sheet of paper 20 cm × 20 cm. It can be used to check the magnetic compass of a boat, aircraft or car.

5.11 The Tables used on a Mercator Star Map

The curves can also be drawn on a Mercator star map projection such as the Philips' *Stars of the Middle Heavens* as shown in Fig. 5.9.

In this type of projection the Right Ascensions are shown as parallel vertical straight lines uniformly spaced. The declinations are also straight lines, but are not uniformly spaced, but follow the mathematical relation

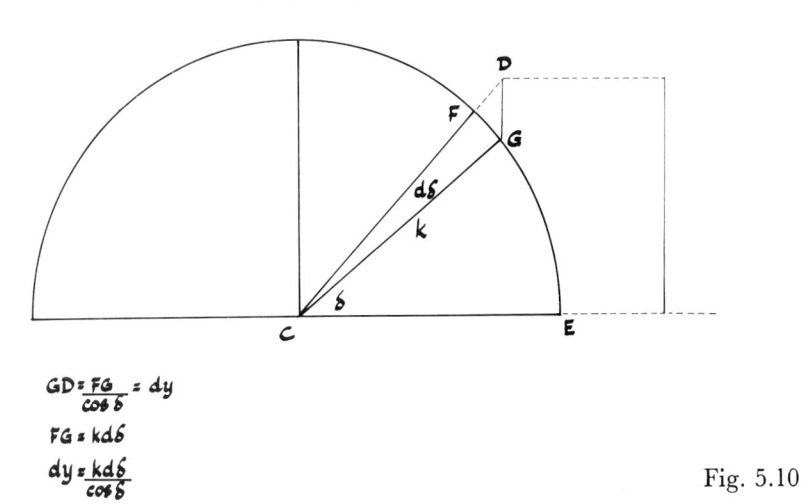

$$GD = \frac{FG}{\cos \delta} = dy$$

$$FG = k\,d\delta$$

$$dy = \frac{k\,d\delta}{\cos \delta}$$

Fig. 5.10

derived from the distance separating two parallels of latitude on a nautical chart, or two parallels of declination on a star chart.

To construct a Mercator star chart in a few minutes with the help of a calculator provides a little instruction and practice, particularly in the use of the key for $\log_e$ ((ln) on some calculators). The method is similar to that used in drawing the navigator's Mercator chart (Section 4.16) but adapted to the celestial sphere. The problem is to find the y coordinates of the declination lines with $y = 0$ as the equator. In Fig. 5.10, GF is the small change in declination, $d\delta$, which appears on a Mercator chart at GD along the y axis on a Mercator projection. $GD = FG/(\cos \delta) = dy$. A small increase in δ then appears as $dy = k\,d\delta \sec \delta$ where k is a constant which depends on the scale we chose to use for our chart.

The full Mercator chart distance EF or

$$y = k \int_0^\delta \sec \delta \, d\delta$$

On the Philips' *Stars of the Middle Heavens*, 15° at the equator measures 45 mm so using this value we find that approximately for a declination of 15° we have

$$45 \text{ mm} = k \int_0^\delta \sec \delta \, d\delta \quad \text{(compare with Section 4.16)}$$

$$= k \log_e \tan\left(45 + \frac{\delta}{2}\right)$$

$$= k \log_e \tan(45 + 7.5).$$

The scientific calculator has a key dealing instantly with logs to the base e or to base 10

$$k = \frac{45}{\log_e(\tan 52.5)} = 169.94$$

and the results follow—each taking but a few seconds to calculate.

The declination line for:
 15° = 169.94 $\log_e(\tan 52.5)$ = 45 mm from the equator.

The declination line for:
 20° = 169.94 $\log_e(\tan 55)$ = 60.56 mm from the equator.

The declination line for:
 30° = 169.94 $\log_e(\tan 60)$ = 93.34 mm from the equator.

The declination line for:
 40° = 169.94 $\log_e(\tan 65)$ = 129.64 mm from the equator.

The declination line for:
 50° = 169.94 $\log_e(\tan 70)$ = 171.75 mm from the equator.

This formula enables one to draw a Mercator star chart of one's own dimensions and the exercise is included here to show that these seemingly awkward calculations are made light of with a calculator. It will be appreciated that stars with declinations near the pole will have a large

Altitude and Azimuth Lines

y ordinate; for example Polaris with declination 89°.166 will have a y ordinate $= 169.94 \log_e(\tan 45 + 44°.583) = 831.376$ mm, so you will need a large piece of paper to include Polaris and its neighbours round the pole! The pole itself cannot be marked on this projection.

5.12 Having prepared your Mercator coordinate system, with hour angles along the x axis and the y axis scaled as described above, the points from the tables for the azimuth lines and the altitude lines can be plotted, and on joining them up in families, the result shown in Fig. 5.9 will be achieved.

From the Fig. 5.9 which shows the azimuth and declination curves drawn on a sheet of acetate overlaying the chart, and made so that they can slide to the east or west according to the time of day and year, the azimuth and altitude of any visible star can be determined.

Interpolation between the curves should not lead to any great inaccuracies, but more curves can easily be calculated and inserted, if required, to reduce inaccuracies from this source.

Setting the planisphere as described in Section 2.6 is equivalent to setting the N–S line or the noon line of the overlay on the RA equal to the local sidereal time at the time of observing. The N–S line shows what star is culminating at the time, and when a star culminates or is on the N–S meridian, the local sidereal time is identical with the star's Right Ascension (see Fig. 2.7). The sidereal time is found either by calculation, as shown in Section 2.8 or directly as given in the *Nautical Almanac* for each day of the year and hour of the day.

Note that the horizon curve in Section 5.11 is of a sinusoidal nature and is of interest in that it indicates the limits of declinations between which stars will be able to rise and set for the latitude of the overlay, i.e., the altitude–azimuth graticule which as we saw is applicable to one particular latitude. The Mercator star map, unlike the polar planisphere is not restricted to any particular latitude, as it does not show its zenith or horizon, although the graticule has to be drawn for each latitude or band of latitudes of a few degrees width.

5.13 ## The Planisphere used as a Sun Compass by Means of the Alt–Azimuth Graticule

The Philips' polar planisphere can be used in conjunction with azimuth–altitude overlay curves to convert it to an accurate sun compass.

Positional Astronomy and Astro-Navigation Made Easy

Fig. 5.11

Set the time on the date in the ordinary way and place a ruler along the line joining the centre of the planisphere to the date on the planisphere and note the point where this line cuts the dotted ecliptic 'ellipse'. This point is where the *mean sun* is at that GMT and date, and where this point lies in the azimuth family of curves will give us some idea of the azimuth of the Sun. This however is not good enough as we are really concerned with the *true sun* for our compass bearing. So we should set against the date not GMT but the true sun's local time which, as we have seen, is $UT + E^{-\text{long W}}_{+\text{long E}}$. ($E$ is the equation of time and is given daily in *Whitaker's Almanack*.

Read the azimuth of the true sun from the azimuth curves and use it to orientate a compass rose as shown in Fig. 5.11. M is a vertical match at the centre and the shadow should fall on the supplement of the azimuth, i.e., (Az) ± 180.

Figure 5.12 shows a sun compass based on the Philips' planisphere, but which has a circular white mask of tracing paper that has inscribed on it the 'ecliptic ellipse' with the mean sun in black on the line joining the date and the pole or centre. It also has marked the position of the true sun as determined by the actual Right Ascension of the Sun.

On some dates, notably in October and November, the true sun is up to

Altitude and Azimuth Lines

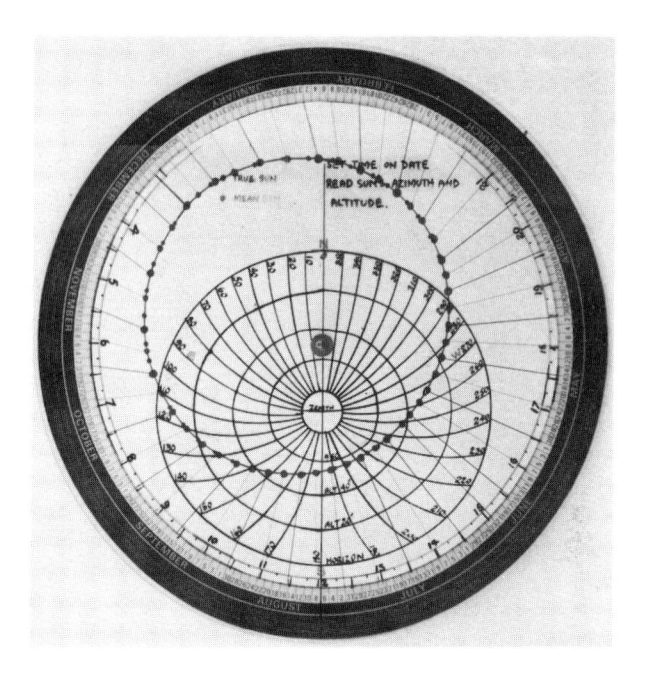

Fig. 5.12. *A Sun Compass.* The positions in (HA) and declination of both the true sun and the mean sun are marked on the ecliptic of the planisphere throughout the year. The overlay of altitude and azimuth curves then shows the Sun's altitude and azimuth at any time of the day. The azimuth indicated can be transferred to the compass rose using a vertical style, a match, which should cast a shadow on the (Az)$\pm 180°$. When this is so, the compass is correctly set and can be used to correct a magnetic compass (see Fig. 5.11).

16 minutes ahead of mean time, and in February it is nearly 15 minutes behind. Nevertheless these true sun lines, although they appear to be offset from the centre, do indicate the position of the true sun and the true azimuth of the Sun. Of course the longitude West or East must additionally be taken into account; that is the user must work with local sun time.

With care the Sun's azimuth can be found to within a degree or two, and so makes a good means of checking a magnetic compass which is liable to errors from deviation and variation.

5.14 The Watch Compass

Many people accept the 'watch rule' for finding the true north by the Sun, because it sounds technical and plausible, but it can lead people off their course and is a menace except within very narrow limits. The 'rule' instructs the traveller to place his watch horizontally and turn it round till the hour hand points in the direction of the Sun's azimuth. Due South is then allegedly given by the line midway between the hour hand and the 12.00 mark on the watch face. Our calculator can tell us what really happens and we can use it to make an accurate sun compass.

The idea of dividing the angle between the Sun and the watch hand is plausible because the watch hand travels in angular motion twice as fast as the Sun in its path, but unfortunately the paths are not in the same plane! So the Sun's hour angle (which measures sun time) and the Sun's azimuth, which is what the watch hand attempts to measure, do not keep in step—as our spherical triangles show, and as the table in Section 5.16 shows.

The calculator can show us precisely what happens. For example: on 22nd August, Sun's declination 20°.2 you are by your map in lat 53°.4. Your watch says LMT $15^h\ 15^m$. The hour hand of the watch has turned through 97°.5 from 12 o'clock (30° each hour) and would show the Sun's azimuth as $97°.5/2 = 48°.75$ from the direction the Sun was at 12 o'clock.

At $15^h\ 15^m$ the Sun's local hour angle is approximately $3^h\ 15^m$ or $(3^h\ 15^m \times 15)°$ or 48°.75.

This is the angle that would be given by the watch hand as the Sun's false *azimuth* west of south. The calculator gives the correct azimuth making use of the Sun's hour angle, the latitude and the Sun's declination, as follows:

$$\cot(Az) = \frac{\cos(HA) \sin \phi - \cos \phi \tan \delta}{\sin(HA)}$$

$$\cot(Az) = \frac{\cos 48°.75 \sin 53.4 - \cos 53°.4 \tan 20°.2}{\sin 48°.75}.$$

The calculator gives, $(Az) = 67°.59$ west of south. The watch method gave 48°.75, so the error would be 19° and this could be serious and even disastrous for a hiker or yachtsman.

5.15. Finding the True North by the Sun Using a Simple Table

All we need to get the azimuth of the Sun using our watch is to have a ready-to-hand simple conversion table (Table 5.12), and to save the bother of guessing half the angle between the big hand and the 12, we should use our watch for its proper purpose—to tell the time!

It is quite unnecessary and dangerously misleading to use your watch face for anything except to tell the time. From the time (GMT) calculate or look up the local sun time.

Altitude and Azimuth Lines

$$\text{Local sun time} = \text{GMT} + E{}^{-W}_{+E} \text{ Long}$$

Then look up the Sun's bearing from the simple table (Table 5.12) shown for the local sun time and the declination. The table can be written on a postcard. Use this bearing to set the compass rose as shown in Fig. 5.11. The column on the right shows the azimuth deduced from the 'watch trick'. Compare with the true azimuths. At midday there is no discrepancy.

This simple device for using the Sun's correct azimuth to orientate a compass can be further simplified by drilling a small hole in the centre of a 360° protractor and sticking a match or toothpick in it vertically and turning the protractor in a horizontal plane until the shadow falls on the supplement of the Az.

Useful calculated figures for the azimuth of the Sun at various times of the day and at different times of the year. Interpolate for intermediate values for δ and for local sun times.

The right hand column shows the azimuth indicated if the 'watch' setting is relied upon. Errors are a few degrees in mid-winter, but over 20° in the summer when the Sun is most likely to be used by travellers.

Table 5.12. Azimuths of Sun in degrees.

Local Sun Time	June δ+23°	July May δ+20°	April August δ+10°	March Sept δ=0°	Oct Feb δ−10°	Nov Jan δ−20°	Dec δ−23°	Watch result
4	53							60°
5	64	66						75°
6	75	77	84	90				90°
7	86	88	95	102	108			105°
8	95	100	108	114	120	125	127	120°
9	112	114	122	128	133	138	139	135°
10	129	132	138	143	147	151	152	150°
11	152	154	158	161	163	165	166	165°
12	180	180	180	180	180	180	180	180°
13	208	206	202	199	197	195	194	195°
14	231	229	222	217	213	209	208	210°
15	248	246	238	232	227	220	221	225°
16	262	260	252	246	240	235	233	240°
17	274	272	265	258	252			255°
18	285	283	276	270				270°
19	296	294						285°
20	307							300°

Fig. 5.13

Expression used for Azimuth:

$$\cot(Az) = \frac{\tan \delta \cos \phi - \sin \phi \cos(HA)}{\sin(HA)}$$

HA is the LHA$^\circ$ = GMT + 12 + E^{-W}_{+E} Long

use this relation to avoid any error that may arise from interpolation, or check the Azimuth from the curves in Fig. 5.7.

5.16 Figure 5.13 shows a sun compass made from a 360° protractor suspended in a cradle consisting of a strip of aluminium which has a fine silt. The protractor can rotate in its cradle and is set for the plane of the slit to lie on the azimuth read from Table 5.12. Turn the whole device until the slit of light made by the Sun falls on the centre of the protractor (the axis of rotation). Then the protractor is correctly orientated. The cradle can be suspended by a fine string so that the plane of the protractor is horizontal. In a wind a small spirit level can be used to ensure that the protractor is in a horizontal plane.

Altitude and Azimuth Lines

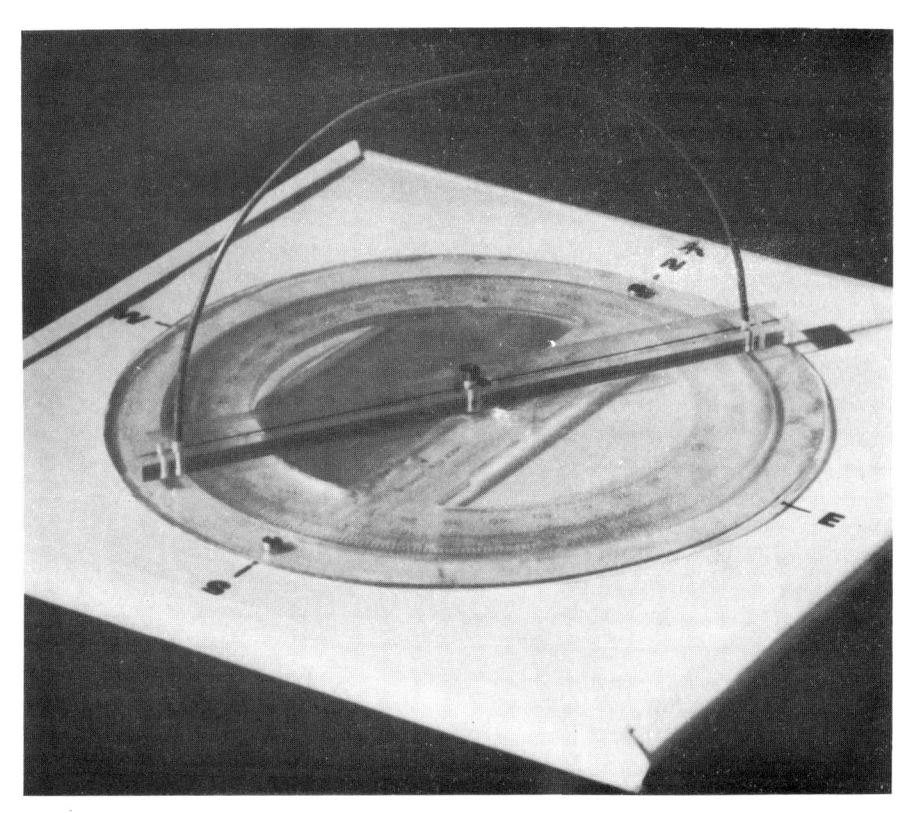

Fig. 5.14

An alternative pattern for a sun compass, the Sun's azimuth, is shown in Fig. 5.14. It is really a refinement on the matchstick compass of Fig. 5.11.

A piece of stout wire is bent in the form of a semi-circle and is fixed to a cross-piece that is free to turn about an axis at the centre of a 360° protractor. The cross-piece is turned so that the wire casts a single line shadow across the centre of the protractor, which is then turned so that the shadow is in the Sun's azimuth as found from Table 5.12. The protractor is then truly orientated (see Section 5.15.)

5.17 We have seen that the values calculated and shown in the tables of Sections 5.6 and 5.9 of this chapter can be used to produce azimuth and altitude curves as planisphere transparent overlays for three different kinds of star maps having (1) polar coordinates, (2) rectangular coordinates, and (3) Mercator projection chart coordinates.

Perhaps the most interesting and possibly unpublished application of these tables is to the mediaeval astrolabe which has a stereographic projection star map. This will be considered in the next chapter.

145

6 The Mediaeval Astrolabe. A New Approach

Positional Astronomy and Astro-Navigation Made Easy

6.1 History

The numerical solution of appropriate spherical triangles is one of the important tasks in positional astronomy and in astro-navigation. It was in effect the main function of the mediaeval astrolabe to perform this task by an ingenious use of geometry, but it was not all child's play despite Chaucer's essay on the subject for his 'Sone Litell Lowys aged 10'. As we have seen in Section 3.1, the story of spherical triangles and related calculations goes back some 2500 years to Pythagoras.

The astrolabe was the ingenious answer of the early Alexandrian astronomers to the problem of how best to represent on a map the positions of stars as they move across the sky, at a time when spherical trigonometry was practically unknown (see Section 3.1). The solution was a star map by stereographic projection, the principle of which is best explained by a diagram (Fig. 6.1).

6.2

This is the way the early Greek astronomers came to terms with spherical triangles, using plane geometry and so constructed star maps for practical use, by stereographic projection. This was done by taking a South Pole view of the celestial sphere, and projecting stars on to an equatorial plane as in Fig. 6.1. This type of projection was the basis of most astrolabes. The results were plotted by geometrical construction.

An *astrolabe* has two sides. Figure 6.2(a) shows the 'almanac' side showing the signs of the Zodiac. It is fitted with a sighting bar for finding the altitude of the Sun or of a star or planet. Altitudes were then of funda-

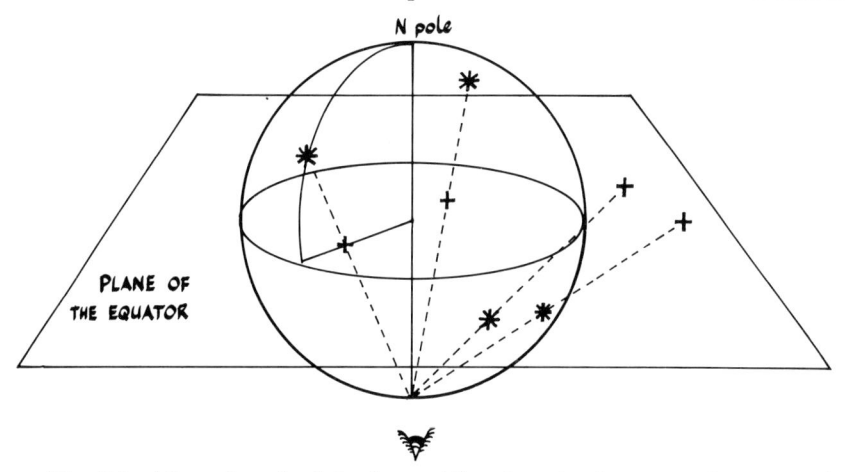

Fig. 6.1. *Stars in celestial sphere. $^+$Stars' projection on to the equatorial plane.

The Mediaeval Astrolabe

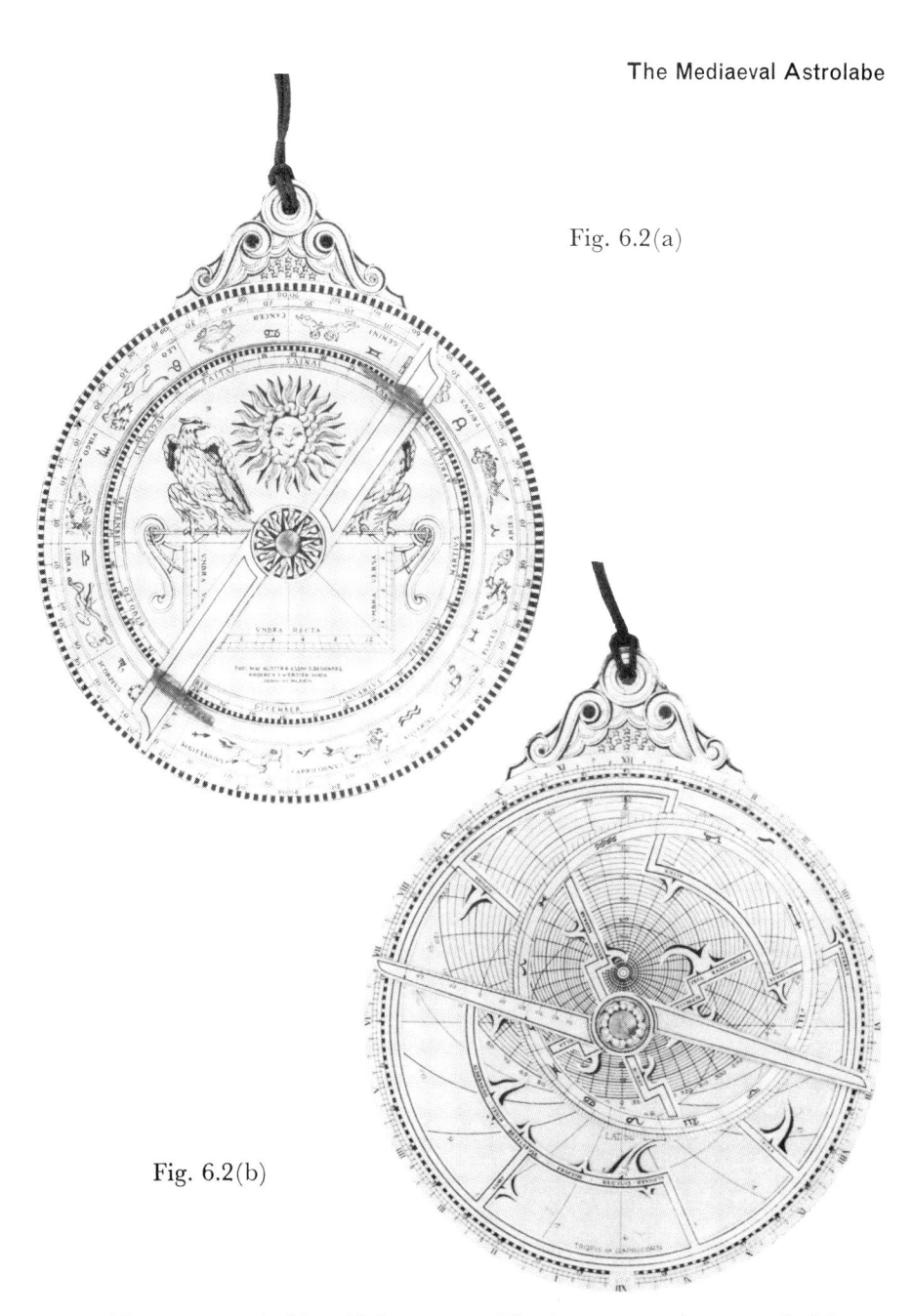

Fig. 6.2(a)

Fig. 6.2(b)

mental importance in identifying a star. The instrument is suspended from a ring and hangs vertically when taking an altitude. The almanac information and the measured altitude are then transferred to the other side of the astrolabe (Fig. 6.2(b)), which is then set to incorporate the observed altitude and date in a manner analogous to the setting of the planisphere described in Section 2.4.

This side is the 'calculator' side and in effect solves spherical triangles, *not* electronically, but graphically, using the altitude measured. The astrolabe was the most important tool of the astronomer or traveller until the end of the 17th Century. It became the clock, giving devout Muslims the times for prayer, a compass for a bearing on Mecca, and a planisphere for finding the positions of stars.

6.3 The Mediaeval Astrolabe as an Alt–Azimuth Planisphere

Despite the distinguished and romantic history associated with the mediaeval astrolabe and its craftsmanship, it can do no more in effect in positional astronomy than a good polar planisphere fitted with an accurate overlay of azimuth and altitude curves, as shown in Fig. 5.3. There are two differences.

(1) The polar planisphere as modified in Fig. 5.3 is based on a star map using polar coordinates with Right Ascensions and declinations radiating from the pole as origin. The azimuth and altitude lines make a strange but pleasing pattern.

The star map of the mediaeval astrolabe is formed by a stereographic projection which has the intriguing requirement that both the azimuth and the altitude curves form two separate families of circles or parts of circles. These two families of circles were engraved on a plate called a tympan. Further, the author, while experimenting with these curves made the obvious discovery that if the values given in the tables of Section 5.6 are plotted on the stereographic star chart for a particular latitude, then a perfect tympan results, without recourse to complex graphical constructions. The azimuth lines and the altitude lines form perfect circles which can easily be drawn.

(2) The other difference is largely one of construction. The transparent overlay azimuth and altitude curves used in Fig. 5.3 are easily produced in this age of plastics and polythene sheets, but to the ancient craftsmen who worked in metal, the best solution to the problem of making a star map or rete (Fig. 6.2(b)) that could turn about the tympan and yet show the stars, was to use an openwork structure and mark the stars on it by little pointed projections (see Fig. 6.2(b)) thus overcoming the transparency problem.

A 'rule' lies centrally over the rete and is used to line up the date on the ecliptic circle and the time of day on the hour circle. This operation is the equivalent of 'setting the date on the time' in the Philips' planisphere, as described in Section 2.4.

The Mediaeval Astrolabe

6.4 Two New Approaches to Marking an Astrolabe

Many books have been written about the astrolabe (see Bibliography) and most contain instructions on the geometrical construction of the tympan with its azimuth and altitude lines, as well as instructions on the making of the rete or star map by stereographic projection. Strictly speaking each latitude should have its own tympan, although, as with a Philips' planisphere, a tympan has a useful zone of about 5° on either side of the latitude for which it was designed.

The following sections describe a new approach to the making of an astrolabe using the calculator. Two methods are considered:

(1) By calculating the radii and the positions of the centres of the circles involved in the azimuth and altitude circles.

(2) By simply plotting the calculated values tabulated in Sections 5.6 and 5.9 on suitably scaled polar graph paper to form these circles.

6.5 The 'Calculated' Astrolabe

In view of the complexity of the geometrical construction of the astrolabe it might appear that to attempt to calculate the positions of all the lines, the azimuths and almucantars would be unrewarding, but with a calculator this is not so, particularly as the azimuth lines are parts of circles and the almucantars are complete circles. Therefore we can dispense with geometry and by means of the calculator reduce the geometry to numbers and simply calculate the radius of each circle and the position of its centre.

6.6 The Azimuth Circles

In Fig. 6.3 are shown the stereographic projections of the zenith A, and the nadir B, which are the two points lying on all the azimuth circles. C, the mid-point between A and B is therefore the centre for one of the azimuth circles. A line DCE drawn at right angles to AB through C must contain the centres of all the azimuth circles as they must all pass through A and B.

The circle with its centre at C, radius CA will give the azimuths 90 and 270 (when the 90° azimuth passes through the zenith it becomes Az 270)

Now mark points F, G, H, etc. such that $CAF = 10°$, $FAG = 10°$, etc., then F will be the centre of the common azimuth circle for both 80° and 260°, G the centre of the common azimuth circle for 70° and 250° and H for azimuth circles 60° and 240°, and so on.

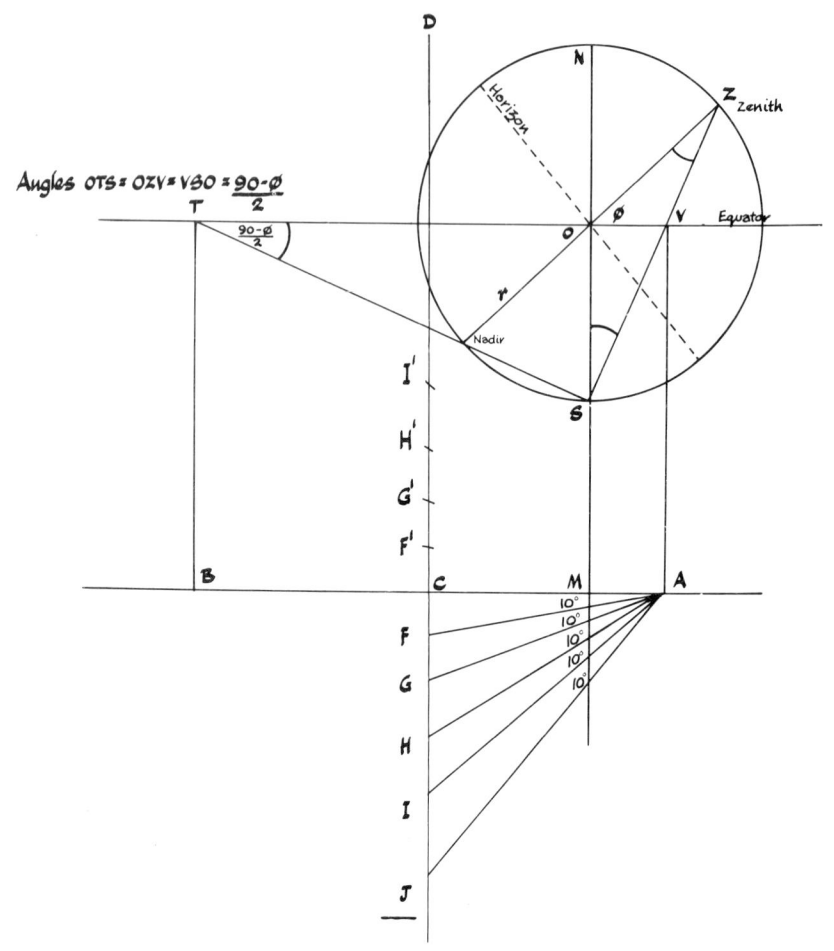

Positional Astronomy and Astro-Navigation Made Easy

Fig. 6.3

6.7 At this point we bring in the calculator. We require the distance AB (for convenience consider the radius of the circle representing the celestial sphere as 60 mm). This consists of $AM+MB$.

From Fig. 6.3, taking the latitude of the place for working as $51°$, then the angles are as marked, and are each $= \dfrac{90-\phi}{2} = 19°.5$.

$$60 \text{ mm}/AM = \tan 19.5 \quad \text{also} \quad MB/60 = \tan 19.5$$

$$BM = 169.43 \text{ mm} \quad MA = 21.247 \quad AB = AM+MB$$

and the required distance $AB = 190.68$ mm.

The points F, G, H, etc., are readily found from

$$CF = CA \tan 10°$$
$$CG = CA \tan 20°$$
$$CH = CA \tan 30°$$
$$CI = CA \tan 40°, \text{ etc.}$$

where $CA = AB/2 = 95.34$.

We are now able to draw all the azimuth circles to any accuracy required (see Table 6.1).

6.8 How to Draw the Azimuth Circles

For the azimuth circles (e.g., latitude 51° and for radius at equator 60 mm) draw $AB = 190.68$ mm and mark the mid point C.

One azimuth circle for (Az) = 180° has been drawn and we now describe circles for eight other circles giving nine azimuth circles.

Draw DE at right angles to AB through C.

Mark points, F, G, H, I, J, K, L, M, N and F', G', H', I', J', K', L', M', N' at distances from C as follows:

Table 6.1

Distance of centre from $C = CA \tan(\text{Az})$			Radii $= \dfrac{CA}{\cos(\text{Az})}$
F and F'	= 16.8 mm	$CA \tan 10$	96.81 mm
G and G'	= 34.7 mm	$CA \tan 20$	101.46 mm
H and H'	= 55.0 mm	$CA \tan 30$	110.19 mm
I and I'	= 80.0 mm	$CA \tan 40$	124.45 mm
J and J'	= 113.6 mm	$CA \tan 50$	148.32 mm
K and K'	= 165.1 mm	$CA \tan 60$	190.68 mm
L and L'	= 261.8 mm	$CA \tan 70$	278.75 mm
M and M'	= 540.7 mm	$CA \tan 80$	549.04 mm

6.9

With a compass draw a series of circles with centres at C, F, G, H, I, J, K, L, M, and at F', G', H', I', J', K', L', M', as in the diagram, and radii CA, FA, GA, HA, IA, JA, KA, LA, MA, and $F'A$, $G'A$, $H'A$, $I'A$, $J'A$, $K'A$, $L'A$, $M'A$. The circles are the *azimuth circles*. A is the *zenith of the tympan*, B is the *nadir*. The azimuths are marked in steps of 10°, all pass through

Positional Astronomy and Astro-Navigation Made Easy

A and *B*. The line *AB* is part of a circle of infinite radius and marks azimuths 0° and 180°.

When the full set of azimuth circles has been drawn, a pleasing set of curves is produced, as in Fig. 6.4(a). It is a matter of convention how these are marked. Usually they are marked as shown in the diagram with zero azimuth to the north.

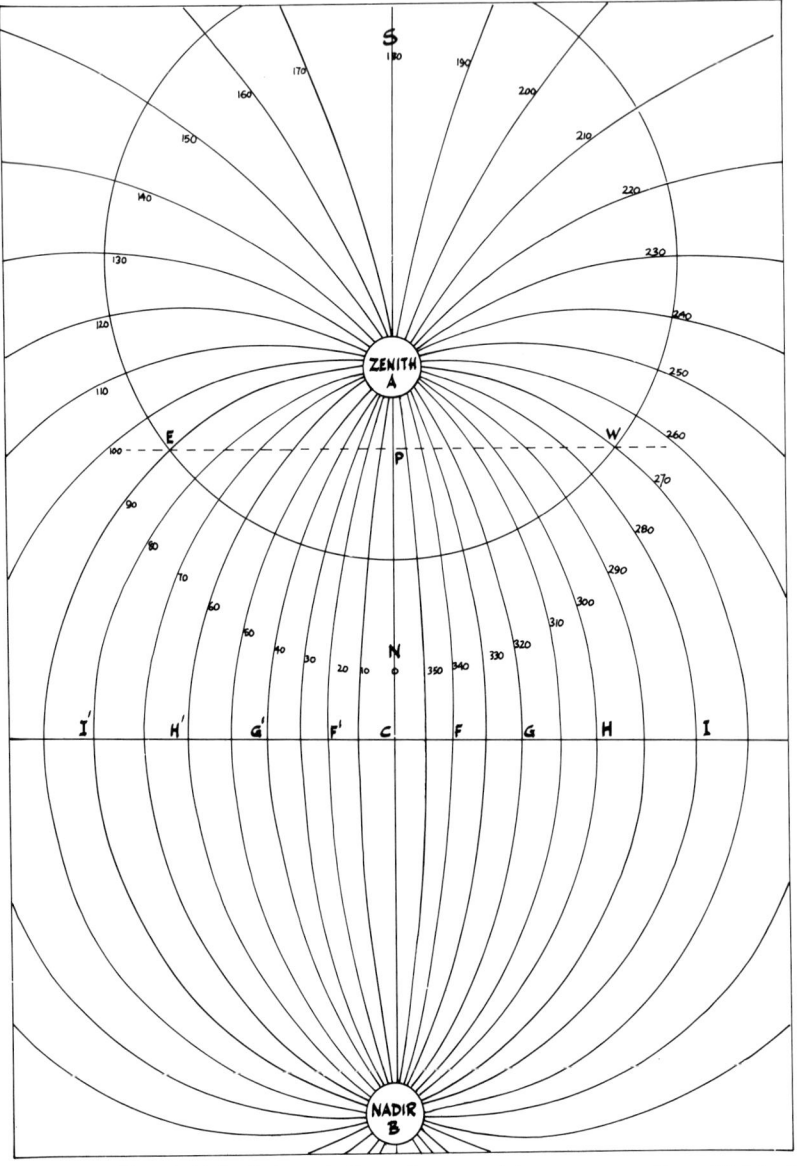

Fig. 6.4(a). Azimuth lines drawn with a compass from Table 6.1.

The Mediaeval Astrolabe

6.10 The tabulated figures giving the centres and the radii of the azimuth circles can be directly derived from trig formulae. Two general formulae can be produced as follows:

In Fig. 6.3, r = radius of the celestial sphere, ϕ is the latitude, BA, the projected distance between the zenith and the nadir, can be considered as
$BM + MA$

$$= TO + OV$$

$$= r \cot\left(\frac{90-\phi}{2}\right) + r \tan\left(\frac{90-\phi}{2}\right)$$

$$= r \left[\cot\left(\frac{90-\phi}{2}\right) + \tan\left(\frac{90-\phi}{2}\right)\right].$$

Using formulae for compound angles this reduces surprisingly and satisfactorily to $2r/(\cos \phi)$.

And for $r = 60$ mm and $\phi = 51°$, $BA = 190.68$ mm, and therefore $CA = 95.34$ mm, as in Section 6.7.

Further, in Section 6.7 the distances CF, CG, CH, CI, etc., are given by

$$CB \tan(90-(Az)) \quad \text{or} \quad CB \cot(Az)$$

but,
$$CB = \frac{r}{\cos \phi}$$

$$= r \sec \phi.$$

Therefore distances from C are given by, $r \cot(Az) \sec \phi$.

The tabulated distances in Section 6.7 for $BA = 190.68$ mm and $\phi = 51$ can be checked by the calculator and the two methods of approach, Sections 6.7 and 6.10, will produce the same table and the same pattern of azimuth circles.

When drawing the azimuth curves it is interesting and instructive to draw complete circles as far as the paper will permit. At least the circle should pass through both the zenith and the nadir as was done when we drew the azimuth curves for the planisphere with polar coordinates (see Section 5.4).

6.11 The Almucantars or Altitude Circles

In order to draw the almucantars or altitude circles we find the diameters of the circles from the fact that a star on the meridian has an altitude given by, (alt) $= 90 - \phi + \delta$, when southing. For a star having declination $> 90 - \phi$ it will have two altitudes, an upper and lower transit, as it will be circumpolar (see Section 10.5).

Positional Astronomy and Astro-Navigation Made Easy

Almucantars or altitude curves for Latitude 51° N or S

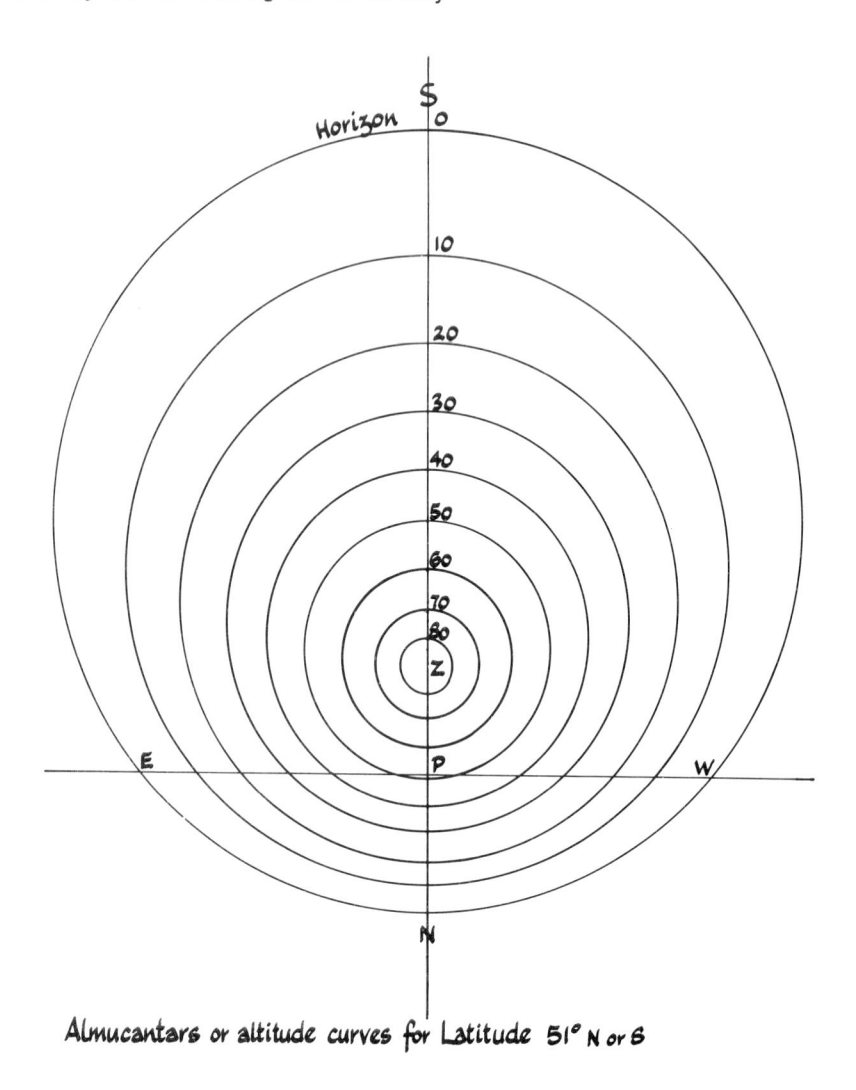

Fig. 6.4(b).

There are thus two declinations which can provide a particular altitude, δ_1 and δ_2. These can be shown by a simple diagram to be,

$$\delta_1 = \phi - 90 + (\text{alt}) \qquad [6.11(1)]$$
$$\delta_2 = 90 - \phi + (\text{alt}). \qquad [6.11(2)]$$

The Mediaeval Astrolabe

On a stereographic projection each declination, δ, on the star chart is at a distance, d_1 from the pole where

$$d_1 = 60 \tan\left(\frac{90-\delta_1}{2}\right).$$

60 mm is the radius of the celestial sphere which is being subjected to the stereographic projection.

The relation is easily established by Fig. 6.5.

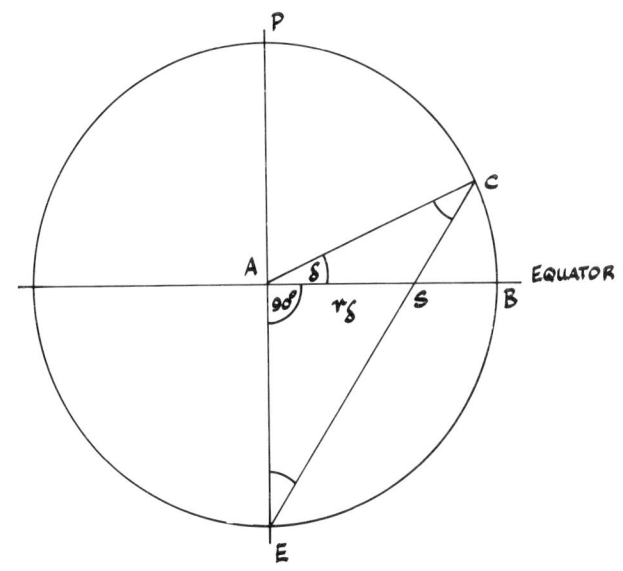

Fig. 6.5. $ACS = AES = \dfrac{90-\delta}{2}$

S is the stereoscopic projections of the star, C, on to the plane of the equator.

$$AS = r_\delta \qquad AC = AB = r_0 \quad \text{(when } \delta = 0\text{)}$$

$$\text{angle } ACS = \frac{90-\delta}{2} = AEC$$

$$r_\delta = r_0 \tan\left(\frac{90-\delta}{2}\right).$$

This is the scale to be used in marking declinations on the astrolabe, i.e., for turning declinations into distances on the tympan (see Fig. 6.7 and Table 6.4).

Table 6.2 gives the pairs of declinations, columns 1 and 2, corresponding to the particular altitudes from 80° to 0° in steps of 10°. Columns 3 and 4 give the distances on the planisphere corresponding to the declination in 1 and 2 using the scaling formula for stereographic projection (Fig. 6.5)

$$d_1 = 60 \tan\left(\frac{90-\delta_1}{2}\right) \quad \text{and} \quad d_2 = 60 \tan\left(\frac{90-\delta_2}{2}\right).$$

Column 5 gives the diameter of the almucantars (d_1-d_2) and column 6 the radius of the circles. The distance of the centre from the pole is given by $\left(d_1 - \frac{d_1-d_2}{2}\right)$.

Table 6.2. All the necessary data for constructing the astrolabe's almucantars (lat. 51°) using only a ruler and compass, namely position of centre and radius from the last two columns, 6 and 7 Distances in mm.
From Fig. 6.3 the distance from the pole P to Z, the zenith (projected) $MA = 60 \tan 19.5 = 21.247$ mm. The centres are at a distance $(d_1+d_2)/2$ from P, and radii are $(d_1-d_2)/2$ with signs as shown.

Altitude	The two declinations on the meridian to give the altitude		Distances from pole $= 60 \tan\left(\frac{90-\delta}{2}\right)$		Diameter of almucantar	Radius of almucantar	Distance of centre from the pole, P
	δ_1	δ_2	for δ_1 d_1	for δ_2 d_2	d_1-d_2	$\frac{d_1-d_2}{2}$	$d_1 - \frac{d_1-d_2}{2}$
	1	2	3	4	5	6	7
80	41°	61°	27.3	15.5	11.8	5.9	21.4
70	31°	71°	33.9	10.04	23.85	11.93	21.97
60	21°	81°	41.2	4.72	36.48	18.24	22.98
50	11°	89°	49.46	−0.52	49.98	24.99	24.47
40	1°	79°	58.96	−5.77	64.73	32.36	26.6
30	−9°	69°	70.25	−11.12	81.37	40.68	29.57
20	−19°	59°	84.12	−16.63	100.75	50.37	33.74
10	−29°	49°	101.85	−22.433	124.3	62.14	39.71
0	−39°	39°	125.80	−28.6	154.4	77.20	48.6

6.12 Altitude Circles by Trigonometrical Formulae

As was done in the case of azimuth circles, we can calculate the radii and the positions of the centres of altitude circles, using a trigonometrical relation again derived from the geometry of the stereographic projection, as illustrated in Fig. 6.6.

ϕ is the latitude, a is the altitude above the horizon.

$$POH = \phi$$
$$EOH = 90-\phi$$
$$EOQ = 90-\phi+a$$
$$QOS = EOQ+90 = 180-\phi+a$$

$$\therefore \quad d_1QO = d_1SO = \frac{180-180+\phi-a}{2}$$

$$= \frac{\phi-a}{2}$$

$$\therefore \quad d_1O = D_1O_1 = r\tan\left(\frac{\phi-a}{2}\right).$$

Also,
$$SOH_1 = \phi$$
$$SOQ_1 = \phi+a$$
$$2OSQ_1 + SOQ_1 = 180$$
$$OSQ_1 = \frac{180-\phi-a}{2}$$

$$d_2O = r\tan\left[90-\left(\frac{\phi+a}{2}\right)\right]$$

or,
$$D_2O_1 = r\cot\left(\frac{\phi+a}{2}\right).$$

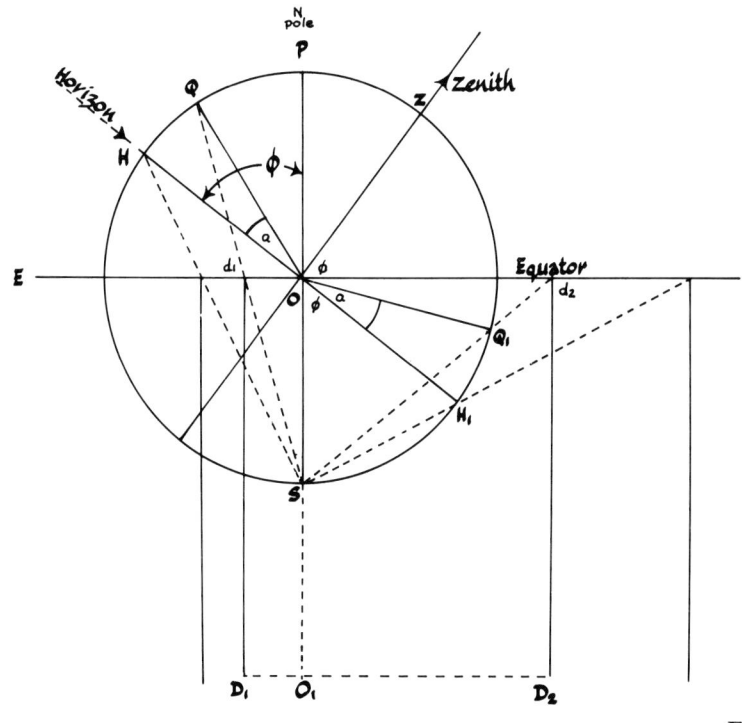

Fig. 6.6

The diameter for the almucantar for altitude $a = D_1O_1 + D_2O_1$

$$= r\left[\tan\left(\frac{\phi-a}{2}\right) + \cot\left(\frac{\phi+a}{2}\right)\right].$$

Then using trigonometrical formulae for compound angles this can be reduced to

$$\frac{2r\cos a}{\sin\phi + \sin a},$$

or radius of circle is

$$\frac{r\cos a}{\sin\phi + \sin a}. \qquad [6.12(1)]$$

Next to find the centre of the almucantar circle.

This will be $(D_1O_1 - D_2O_1)/2$ from O,

i.e.,

$$\frac{r}{2}\left[\tan\left(\frac{\phi-a}{2}\right) - \cot\left(\frac{\phi+a}{2}\right)\right]$$

and this reduces to

$$\frac{r\cos\phi}{\sin\phi + \sin a}. \qquad [6.12(2)]$$

From [6.12(1)] and [6.12(2)] we can, using the calculator, compile columns 6 and 7 respectively in Table 6.2 and so we can draw the almucantars of the astrolabe for latitude ϕ. See Fig. 6.4 showing both azimuth curves and altitude circles.

Table 6.3. The almucantars or 'altitude circles' calculated from the trig formulae above [6.12(1)] and [6.12(2)] are in agreement with Table 6.2. These are not concentric, but can be constructed as follows: R = the radius of the diagramatic celestial sphere of Fig. 6.6.

Altitude	Radius of circle R from $R = \dfrac{60\cos(\text{alt})}{\sin\phi + \sin(\text{alt})}$	Distance of centre from pole from $d = \dfrac{60\cos\phi}{\sin\phi + \sin(\text{alt})}$	
80°	0.59 mm	21.40 mm	
70°	11.90 mm	21.97 mm	Distance of the pole from zenith, A
60°	18.24 mm	22.98 mm	
50°	25.00 mm	24.47 mm	$= R\tan\dfrac{90-\phi}{2}$
40°	32.36 mm	26.60 mm	
30°	40.68 mm	29.57 mm	$= 60\tan 19.5$
20°	50.37 mm	33.74 mm	
10°	62.14 mm	39.71 mm	$= 21.247$ mm
0°	77.20 mm	48.60 mm	

6.13 The Astrolabe Rete or Star Map

This should be drawn on transparent plastic—or alternatively the azimuth–altitude circles should be formed on a transparent overlay.

Stars are mapped by a stereographic projection of the celestial sphere. Simple geometry produces this star map which has a declination scale based on the relation

$$r_\delta = r_0 \tan\left(\frac{90-\delta}{2}\right)$$

where r_δ is the radius of the declination circle δ, and r_0 is radius of equator, $\delta = 0$. The declination circles are shown in Figs. 6.7 and 6.8. The radii and declinations are tabulated below. r_0 is taken to be 60 mm. The astrolabe rete coordinate system is shown in Fig. 6.7.

6.14

In Section 6.13 we saw how the stereoscopic star map could be prepared. It is similar to the ordinary planisphere which is based on polar coordinates, but the declination scale of the stereoscopic map becomes more and more extended as declinations decrease from the North Pole, i.e., from $+90°$ to $0°$ to $-90°$.

The azimuth and altitude curves which were plotted on a polar planisphere, using the data in Tables 5.1 and 5.11 can be plotted on the

Table 6.4 For plotting stars on a stereographic star map as in Fig. 6.7.

Declination δ	$\dfrac{90-\delta}{2}$	$r_\delta = 60 \tan\left(\dfrac{90-\delta}{2}\right)$
90°	0°	0 cms
80°	5°	0.5249
70°	10°	1.058
60°	15°	1.608
50°	20°	2.184
40°	25°	2.800
30°	30°	3.464
20°	35°	4.201
10°	40°	5.035
0°	45°	6.000
−10°	50°	7.150
−20°	55°	8.659
−30°	60°	10.392
−40°	65°	12.867

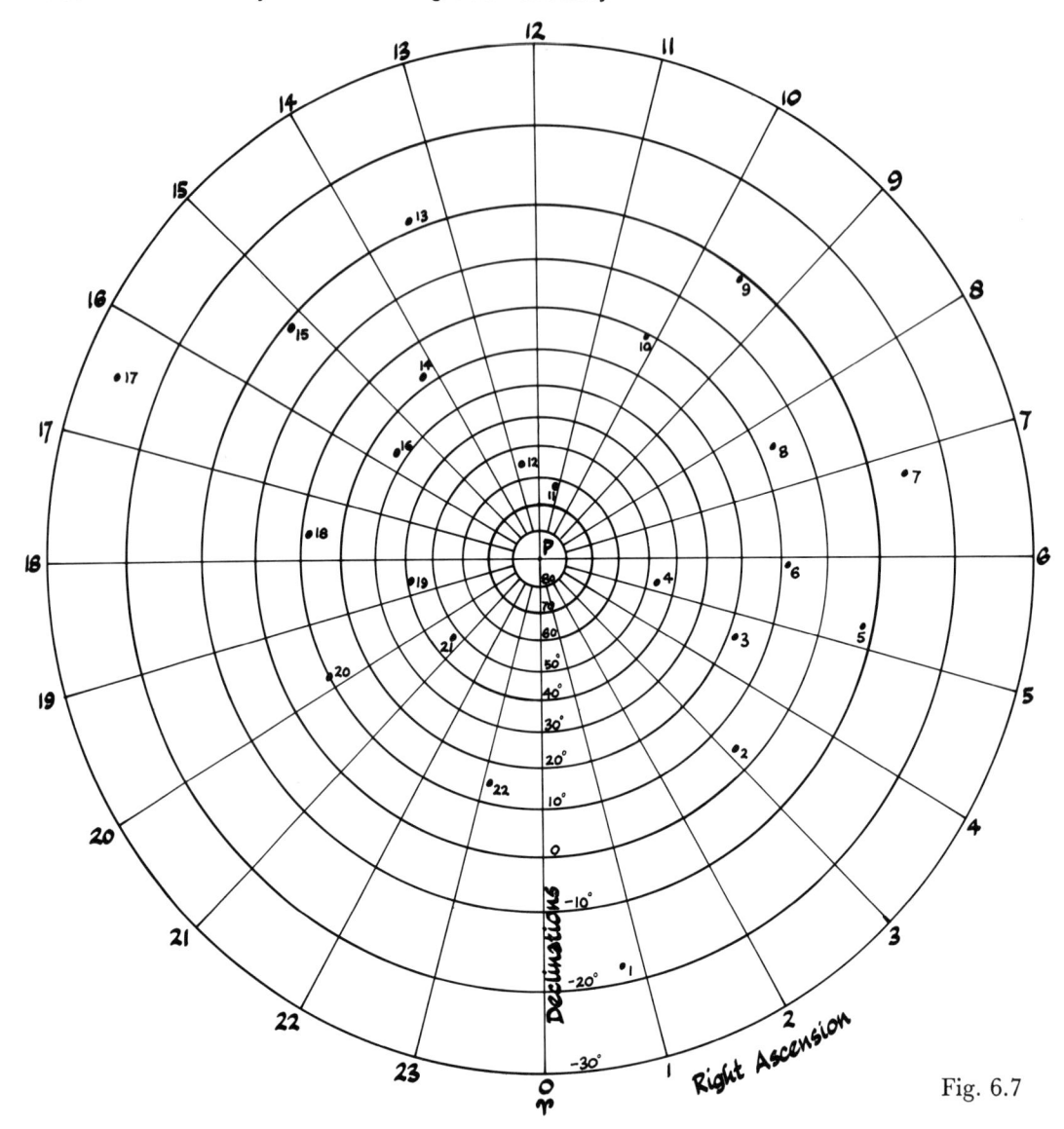

Positional Astronomy and Astro-Navigation Made Easy

Fig. 6.7. Key to the stars numbered on the stereographic star chart
1. Diphda
2. Menkar
3. Aldebaran
4. Capella
5. Rigel
6. Betelgeuse
7. Sirius
8. Procyon
9. Alphard
10. Regulus
11. Dubhe
12. Alioth
13. Spica
14. Arcturus
15. Lirrae
16. Alphecca
17. Antares
18. Rasalhague
19. Vega
20. Altair
21. Deneb
22. Markab

stereoscopic star map using the same data. This plotting produces, as might be anticipated, a set of circles all passing through the zenith and

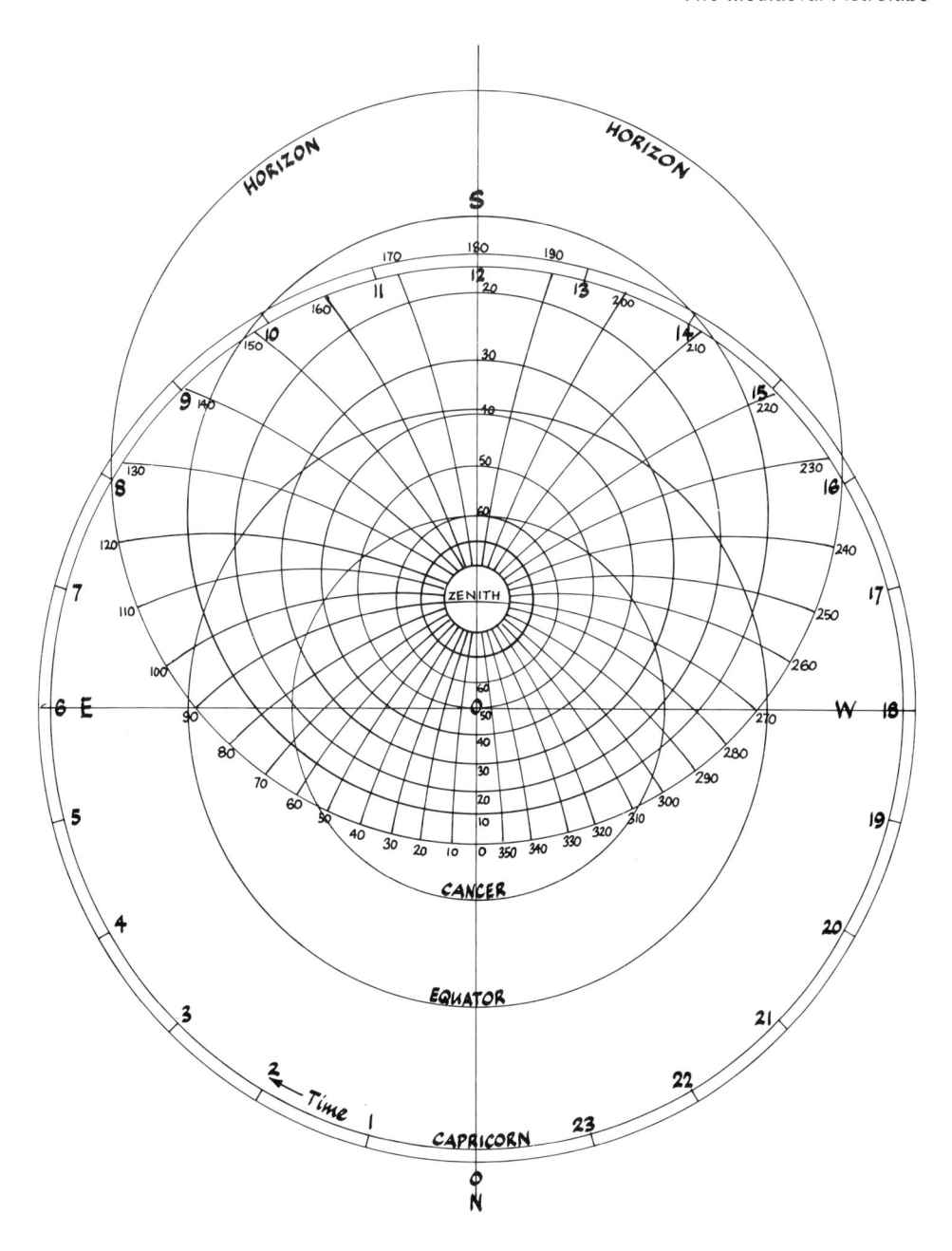

Fig. 6.8 The complete astrolabe is formed by superimposing Fig. 6.4(a) onto Fig. 6.4(b) and both onto Fig. 6.7 giving Figs. 6.8 and 6.9.

nadir of the map for the azimuth lines, and another set of circles for the almucantars.

Tables 5.1 to 5.11 make the marking of a mediaeval astrolabe a matter of plotting the points tabulated. The plotting requires care and is rather tedious, especially near the zenith and the pole, but it is not necessary to plot more than three widely separated points for each circle as these points would theoretically define the radius and the centre of each circle.

The two sets of circles can be shown mathematically to be orthogonal, i.e., to intersect at right angles. This is as might be anticipated, as all altitude lines on a stereoscopic projection are parallel to the horizon and therefore horizontal, and all azimuth lines are drawn from the zenith to the horizon and therefore at right angles to the altitude circles. A property of polar stereographic projections is that angles between stars on the celestial sphere are reproduced without distortion on the projection.

A photograph of an astrolabe drawn by the calculated data in Sections 6.8 and 6.12 is shown in Fig. 6.9.

The azimuth lines and almucantars are on a transparent overlay covering the star map in the same manner as for the planisphere of Fig. 5.3.

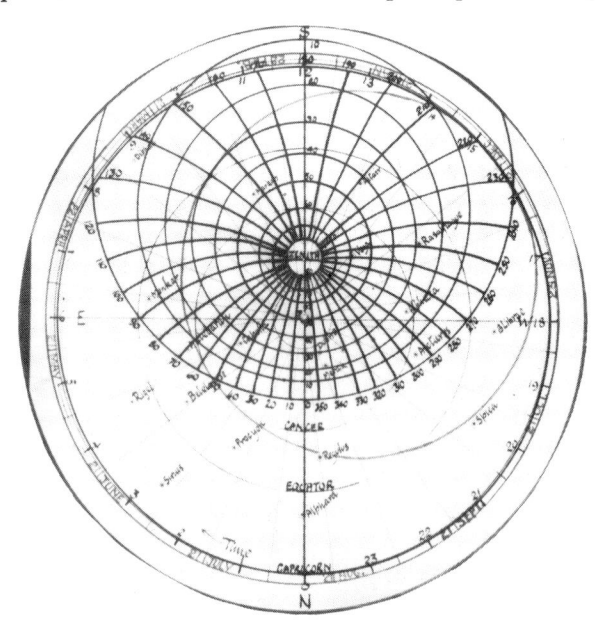

Fig. 6.9

7
Projects with Sundials and the Calculator

7.1 Sundials have always had a special fascination both for astronomers and gardeners from very ancient times. Today they provide a great deal of interest and fun if one accepts and comes to terms with the spherical triangle as a friendly ally rather than a source of confusion and mystery.

The earliest time-measuring instruments known are water clocks, and primitive sundials of ancient Egypt. The first mechanical time-keepers were constructed during the first half of the 14th Century. Watches came much later (1500), but these early clocks and watches were rarely accurate and a sundial was used for checking them.

The old shadow clocks of about 1500 BC were sticks in the sand or vertical pillars or obelisks, but these were not accurate at all seasons, the reason being that the Sun behaves differently at different times of the year.

7.2 *The stick in the sand* idea is not to be despised or discarded as it can be used to find the Sun's altitude by applying the relation,

$$\tan(\text{alt}) = \frac{\text{length of stick}}{\text{length of shadow}} = \frac{AB}{BC} = \frac{H}{L} \quad (\text{Fig. 7.1})$$

after measuring carefully these lengths by means of a steel tape.

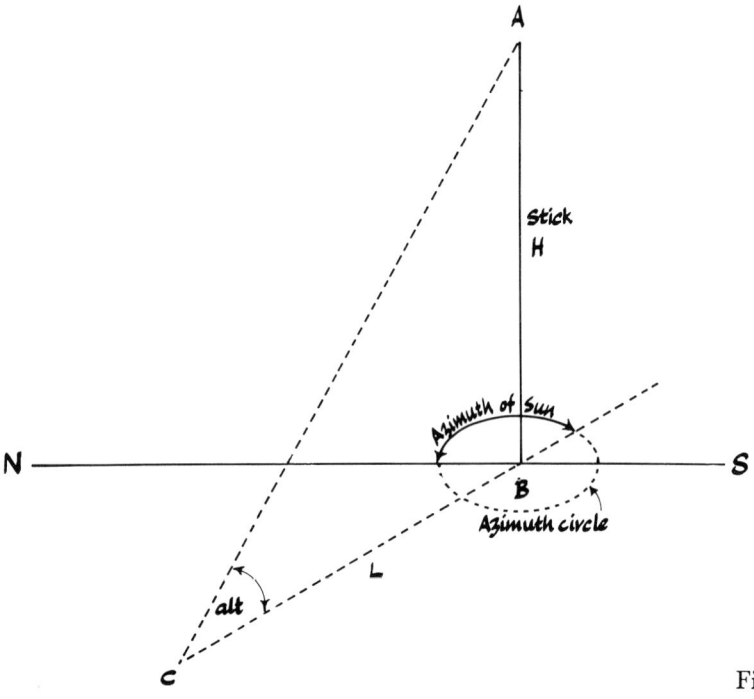

Fig. 7.1

Projects with Sundials and the Calculator

For satisfactory results the 'stick' should be a thin rod, *AB*, of known length, say 60 cm. The surrounding ground should be smooth and level. A plumbline is necessary to ensure that the rod is vertical. An ordinary iron rod retort stand will serve quite well. Surrounding *B* is an azimuth

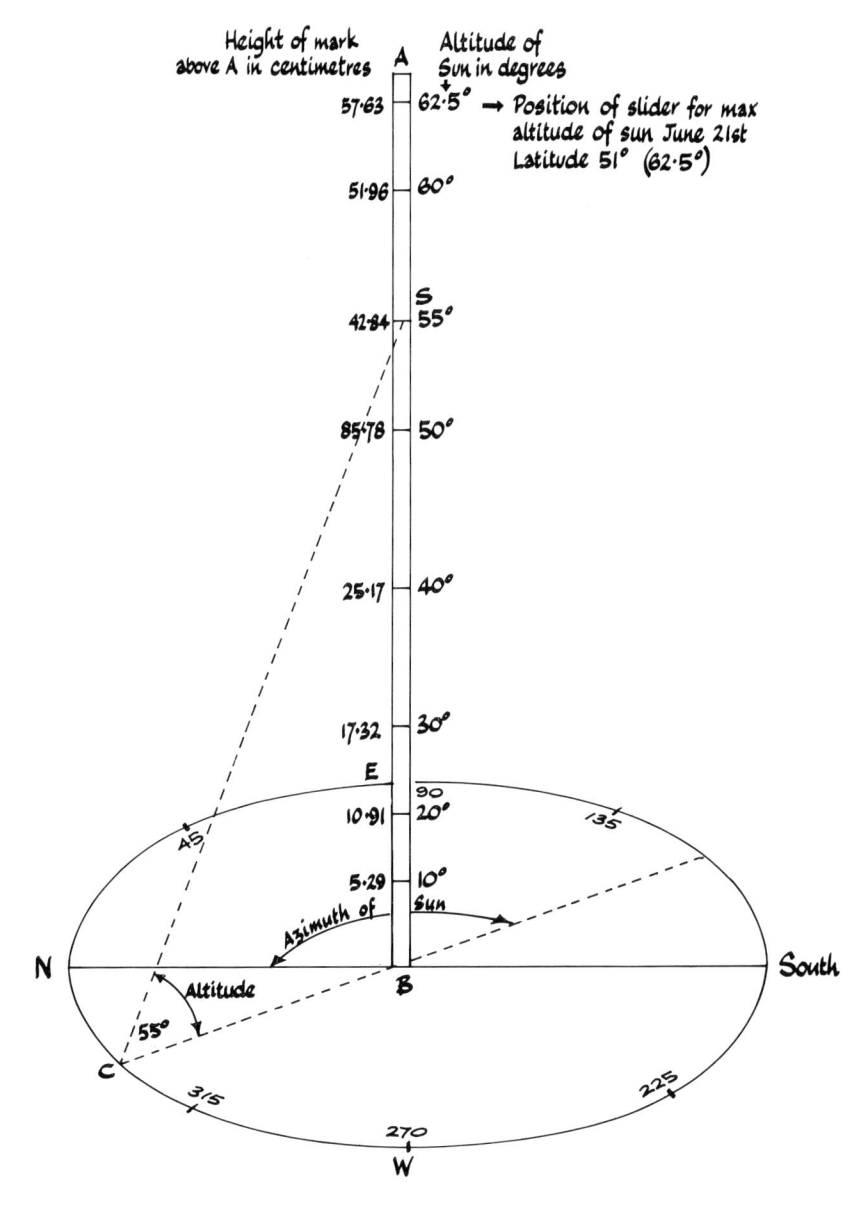

Fig. 7.2. A project sundial described in Section 7.2.

Positional Astronomy and Astro-Navigation Made Easy

circle, such as a 360° blackboard protractor, so that *B* is at the centre of the protractor. This provides a means of measuring azimuths.

A simple variation of this 'stick in the sand' project is shown in Fig. 7.2, and from which the Sun's altitude can be read directly on the calibrated vertical stylus *AB*.

AB is a thin vertical rod about 1 m long at the centre of a horizontal azimuth circle, marked in degrees, and having a clearly defined periphery of radius $CB = 30$ cm. *S* is a small sliding nodus clipped to the rod *AB* and which can move up and down *AB*. The top of this clip should be capable of casting a shadow of the Sun on the circumference of the circle *NESW*. The height of the clip *BS* is related to the altitude of the Sun, *x*, by $\tan x = BS/BC$. From this relation the rod *AB* is then graduated by file marks or paint to read the Sun's altitude *directly*. A few of the positions for latitude 51° are as follows, and intermediate positions can be readily calculated.

Altitude of Sun	Distance *BS*
°	cm.
10	5.29
20	10.91
30	17.32
40	25.17
50	35.78
55	42.84
60	51.96
62.5*	57.63

* This is the maximum position for the Sun at midday on 21st June in latitude 51°.

With these simple devices we are offered a choice of two methods for finding the time. These methods were not available in the early days of astronomy except through complicated geometry or the use of an astrolabic 'calculator' which as we have seen can establish and use the important relations between altitude, azimuths and hour angles.

(1) The Sun's hour angle is given by relation [3.7(1)],

$$\cos(\text{HA}) = \frac{\sin(\text{alt}) - \sin\phi \sin\delta}{\cos\phi \cos\delta}.$$

So if we know our latitude and the Sun's declination we can find the local sun time from the Sun's altitude (see Section 7.11). The azimuth of the Sun is not required.

Projects with Sundials and the Calculator

(2)   The Sun's hour angle is also given by relation [3.7(7)] and by observing the Sun's azimuth on the azimuth circle (the supplement of the shadow mark). We have

$$\sin(\text{Az}) = \frac{\sin(\text{HA}) \cos \delta}{\cos(\text{alt})}$$

from which we can find the Sun's HA, using

$$\sin(\text{HA}) = \frac{\sin(\text{Az}) \cos(\text{alt})}{\cos \delta}$$

which is independent of the latitude.

A number of projects using this very simple vertical rod sundial suggest themselves using the formulae in (1) and (2) involving altitudes, azimuths, hour angles and declinations. No great accuracy can be expected as the shadow tip at low altitudes becomes ill-defined.

7.3   An advance in accuracy and convenience over the vertical stylus was made by the Arabs who set the 'stick' of the shadow clock pointing to the *north celestial pole*, i.e., parallel to the Earth's polar axis. The stylus was thus placed in the N–S meridian, and inclined at an angle equal to the latitude of the place, so that the Sun's hour angles became directly observable.

Books on sundials often gave formulae involving trigonometrical functions, but, with good intentions, expressed them in a form that attempted to facilitate their solution, using log and trig tables. The results were tabulated so that people wishing to make a sundial could mark out the shadow angles. These are however applicable only to one particular latitude, so the reader was left to work out shadow angles relevant to his particular latitude.

The calculator can produce tables of shadow angles in a few minutes for all kinds of sundials and appropriate for the user's latitude. Several examples are given in this chapter.

7.4   There are many types of sundial, but they can be grouped into three classes:

(1)   Those that use a shadow angle that depends mainly and directly on

the Sun's hour angle—as shown by a dial having its stylus pointing to the pole.

(2) Those that depend on the *altitude* of the Sun at a particular time of day and year.

(3) Those that depend upon the Sun's azimuth.

All these can be constructed and marked out with the help of a calculator, but the fascination of making them will be increased if we understand the principles involved in each type. These principles are based on the spherical triangle.

7.5   The simplest to read and one of the most satisfactory type of sundial is the 'equatorial sundial'. A home-made example of this type is shown in Fig. 7.3, and this could be the basis of a school project in metalwork.

It can be made from a chromium-plated cycle wheel sawn into two semi-circular pieces; one supporting the style and mounted on a piece of slate so that the style is in the meridian and pointing to the north celestial pole; the other is fixed at right angles to it in the plane of the equator and is marked in hours. The hour marks are at regular intervals of 15° measured from the semi-circle centre. The hour marks are equi-angular.

In the model shown, a white plastic strip is attached to the inner rim of the equatorial half wheel and is inscribed with the hours as shown. Normally the noon mark is in the meridian but the Sun is not always in the meridian at mean time noon. It may be up to 16 minutes in error because of the 'equation of time' (see Chapter 8). The time strip can however be adjusted on the rim so that the Sun's shadow reads correct GMT or any particular zone time by making an appropriate allowance for the equation of time and for the longitude of the dial.

As the slate base of this dial is horizontal and as the style rises from it at an angle equal to the latitude of the place, the Sun will cast a shadow on the base, so that the device can be marked out as an ordinary horizontal dial, as explained later in Fig. 7.7.

The shadow angles are horizontal projections of the equi-angular shadows cast on the equatorial rim. We are now entering spherical triangle territory, and we shall next deal with the horizontal and vertical dials.

Projects with Sundials and the Calculator

Fig. 7.3. Simple home-made equatorial sundial made from a bicycle wheel rim cut in half. This type can be accurately adjusted to read GMT or BST by adjusting the plastic strip marking the hours. The strip in the photograph shows the mark $12^h\,10^m$ in the meridian which means that when the Sun is in the meridian it is $12^h\,10^m$ local time although the sun time is 12 noon.

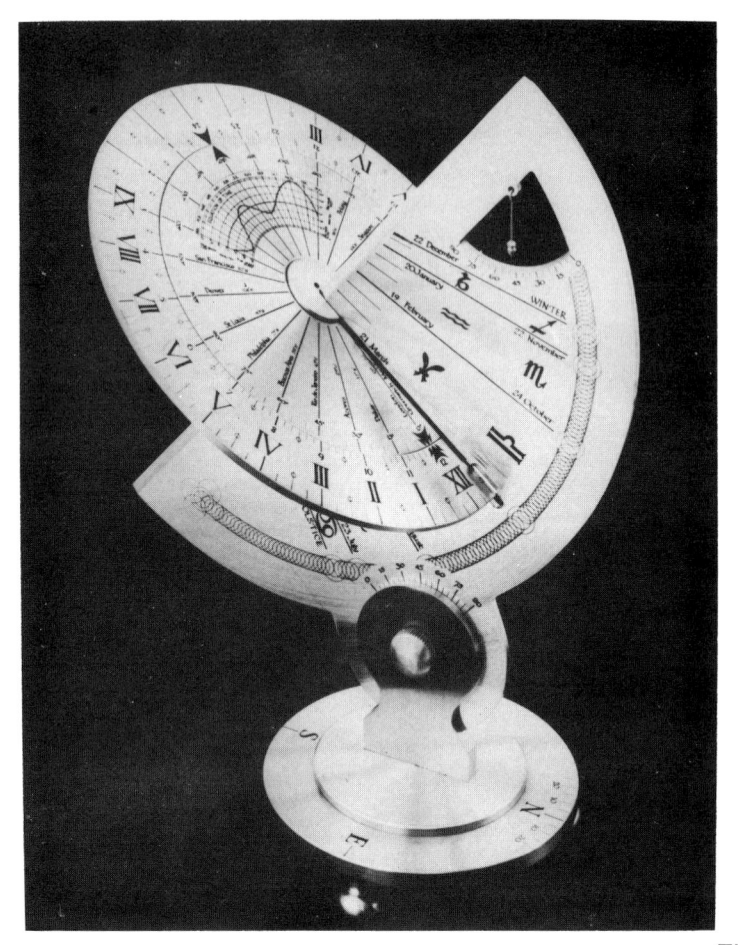

Fig. 7.4

Figure 7.4 shows a modern portable sundial, or 'universal sunclock' by Brookbrae. It is in principle an equatorial dial and is in solid brass. It has a number of interesting features.

(1) An adjustable hour plate which translates local sun time into clock time.
(2) The reverse side of the hour plate records time in winter months.
(3) A revolving plate for telling the time for various longitudes.
(4) A calendar plate showing zodiac months and which records the date by the Sun's shadow.
(5) An equation of time graph.
(6) Adjustable for any required latitude.
(7) Adjustable base feet and plumb check line.

Projects with Sundials and the Calculator

7.6 The Horizontal Dial—
The kind usually found in gardens on a small pedestal

In Fig. 7.5 OP is the style pointing to the pole, P. PNS is the NS meridian, NPT is the hour angle, the angle the Sun makes with the meridian at the pole, P, and TON is the shadow angle γ, i.e., the angle the shadow makes with the meridian.

We now have a spherical triangle PTN with a right angle PNT.

In this spherical triangle, the four consecutive angles and sides we consider are: (1) NPT (2) NP (3) PNT and (4) NT (see Section 3.12).

We can apply the relationship of [3.7(8)] which takes the form given in Section 3.12, i.e.

$$\cos NP \cos PNT = \sin NP \cot NT - \sin PNT \cot NPT$$
$$\quad (2) \quad\quad (3) \quad\quad\quad (2) \quad\quad (4) \quad\quad\quad (3) \quad\quad (1)$$

in which $\quad PNT = 90°, \quad NP = \phi \quad$ and $\quad NT = \gamma$

since $\quad\begin{cases}\cos PNT = 0\\ \sin PNT = 1\end{cases}$

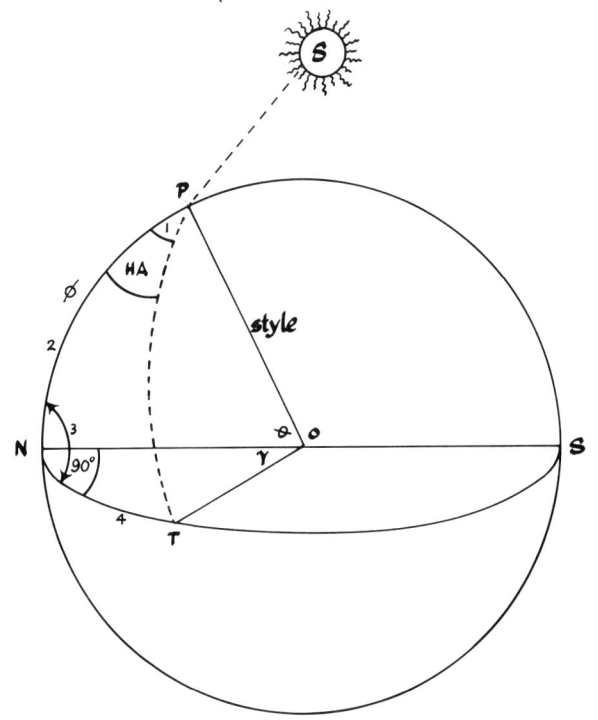

Fig. 7.5. The spherical triangle for a *horizontal sundial* leading to the relation for the shadow angle γ, $\tan \gamma = \sin \phi \tan (\text{HA})$.

Therefore $\qquad 0 = \sin \phi \cot \gamma - \cot(HA)$

∴ $\qquad \tan \gamma = \sin \phi \tan(HA).$

This is the general formula for finding the shadow angle, and is a simple programme for a calculator.

For example, for $\phi = 51°$ and $(HA) = 3\frac{1}{2}$ hrs. $(52°.5)$

$$\tan \gamma = \sin 51 \tan 52.5$$
$$\gamma = 45°.36.$$

Shadow angles corresponding to all possible hour angles can be found in a few seconds, and some are given in the table below in case readers would like to check them.

Fig. 7.6. A simple horizontal sundial made in a school workshop from aluminium sheet.

Table 7.1

Sun time	Hour angle	Shadow angle
12	0	0
11/13	15°	11°45′
10/14	30°	24°10′
9/15	45°	37°51′
8/16	60°	53°23′
7/17	75°	71°
6/18	90°	90°
5/19	105°	109°00′
4/20	120°	126°32′
3/21	135°	142°09′

Projects with Sundials and the Calculator

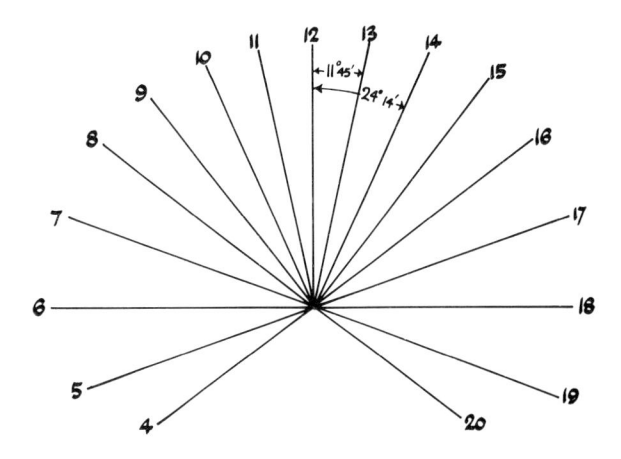

Fig. 7.7. Shadow angles for horizontal sundial. Latitude 51°.

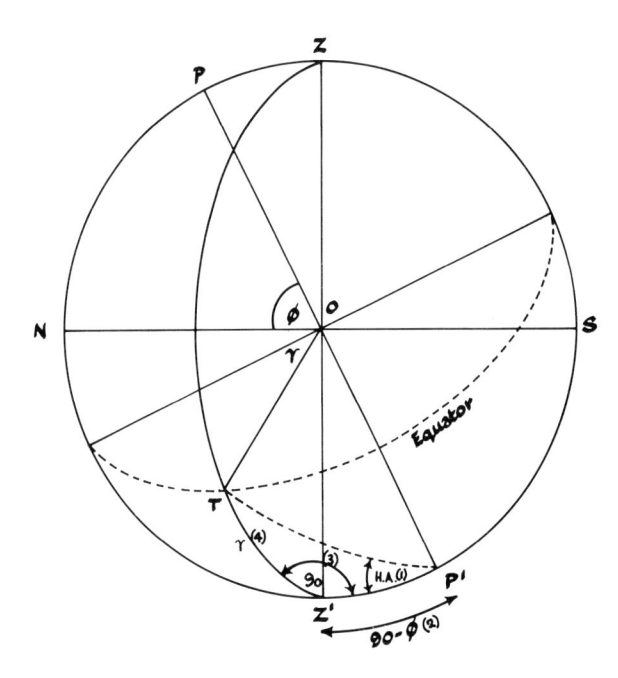

Fig. 7.8

7.7 *In a south facing vertical dial* the treatment is similar and is shown in Fig. 7.8. Z is the zenith, Z' the nadir, P is the North Pole and P' the South Pole.

The relevant spherical triangle is $Z'P'T$, from which using the 'four part' relation [3.7(8)] and Section 3.12 we have

$$\cos Z'P' \cos TZ'P' = \sin Z'P' \cot Z'T - \sin P'Z'T \cot Z'P'T$$
$$(2)(3)(2)(4)(3)(1)$$

175

Positional Astronomy and Astro-Navigation Made Easy

and since $T Z'P' = 90°$

$$0 = \sin(90-\phi) \cot \gamma - \cot(HA)$$

or $\tan \gamma = \tan(HA) \cos \phi$

which is a result similar to that for a horizontal dial but as might be expected, with $\cos \phi$ in place of $\sin \phi$.

Table 7.2 shows shadow angles for a vertical south facing dial from, $\tan \gamma = \tan(HA) \cos \phi$ (see Fig. 7.9).

This type of sundial was often used on the south wall or over the south door of a church. It cannot give times before 06.00 or after 18.00 as the Sun at these times ceases to shine on a south facing wall.

The formula can nevertheless be continued for hour angles greater than 90°, i.e., for times 5/19 and 4/20, as might be required in the summer months. The resulting angles will however be negative and are drawn as

Table 7.2

Sun time	Hour angle	Shadow angle	
12	0	0	0° 00'
11/13	15	9.57	9° 34'
10/14	30	19.97	19° 58'
9/15	45	32.18	32° 11'
8/16	60	47.47	47° 28'
7/17	75	66.94	66° 56'
6/18	90	90.00	90° 00'

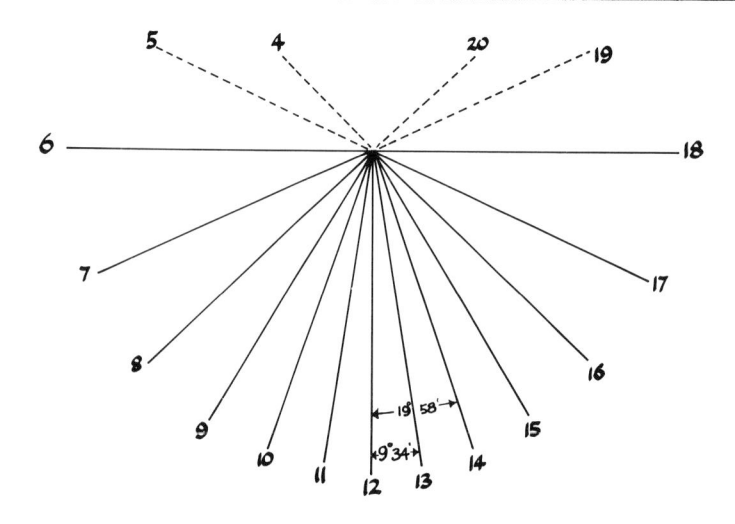

Fig. 7.9. Shadow angles for a south facing vertical dial.

Projects with Sundials and the Calculator

shown, but would appear on the north side of the dial, with the style also on this side of the dial considered as the continuation of the usual dial, still pointing in the direction of the celestial pole, i.e., inclined at an angle 90–ϕ with the dial. Vertical dials can be made to face in any direction, although a south facing dial is the simplest to make.

7.8 The Vertical South Facing Dial—Declining

Many people who are interested in making a sundial to be fitted on to a wall roughly facing south will be frustrated because the wall of the house or the garden does not face due south, but is found to be 40° or 50° out.

A sundial can be marked out to work accurately on such a wall provided the style as always points to the north point of the celestial sphere, and is so set that the shadow of the style falls on the 'vertical noon line' at noon. The markings for the shadow angles for each hour of sun times have to be calculated by a formula which is similar to that for a south facing vertical dial, but has a factor which takes care of the angle which the plane of the wall makes with the east–west line. Further, the style must have a special angle which can be calculated.

The making of sundials affords many opportunities to put to good use the calculator, but first the relations connecting HA's, latitudes and shadow angles have to be established.

7.9

Formula for *the vertical dial displaced by an angle θ from the east–west plane*, so that the wall faces a direction east or west of due south (see Fig. 7.10). Again the spherical triangle is used to find the shape and positioning of the style *SP*. This will be more acute at the top end than for a south facing dial, and it will be displaced from the vertical.

The small Δ *cds* is a spherical triangle, and applying the four part formula as given in Section 7.7:

$$\cos \sigma \cos(90-\theta) = \sin \sigma \cot(90-\phi) - \sin(90-\theta) \cot 90°$$

whence, $\cos \sigma \sin \theta = \sin \sigma \tan \phi$

$$\tan \sigma = \frac{\sin \theta}{\tan \phi} \qquad [7.9(1)]$$

giving σ the angle at which the style should be set from the vertical noon line.

177

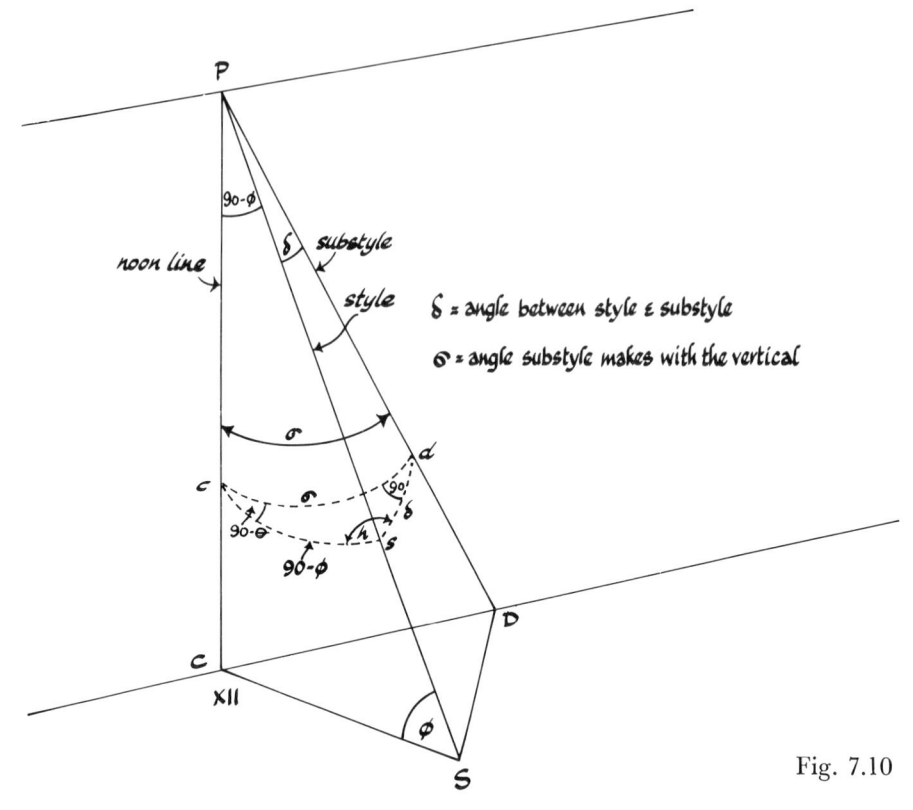

Positional Astronomy and Astro-Navigation Made Easy

Fig. 7.10

δ = angle between style & substyle
σ = angle substyle makes with the vertical

Applying the sine rule in Fig. 3.11,
$$\frac{\sin(90-\theta)}{\sin\delta} = \frac{\sin 90}{\sin(90-\phi)}$$
or $\quad\quad\quad\quad\sin\delta = \cos\phi\cos\theta \quad\quad\quad\quad [7.9(2)]$

giving the angle δ to which the style point is cut.

7.10 An example is taken from a vertical declining dial made for a wall, lat 51°N declining 22° west of south (Fig. 7.11). The style, consisting of a piece of aluminium, was cut to have an angle δ given by relation [7.9(2)], i.e.
$$\sin\delta = \cos\phi\cos\theta$$
whence, $\quad\quad\quad\delta = 35°.7.$

The line *PD* was drawn at an angle σ from the noon line given by,
$$\tan\sigma = \frac{\sin\theta}{\tan\phi} \quad\text{from relation [7.9(1)].}$$
$\sigma = 16°.87$, or 17° for practical purposes.

178

Projects with Sundials and the Calculator

Shadow angles are calculated and measured from PD, the sub-style line.

A cardboard model or Fig. 7.10 will show that the plane of the style PSD, which is perpendicular to the plane of the dial, makes an angle dsc with the plane of the meridian through PC.

This angle, which we shall call h, will be added to or subtracted from the Sun's hour angle when calculating shadow angles. The angle h can be regarded as compensating the normal hour angle on a south facing vertical dial, for the fact that the dial is declining by an angle θ to the east or west of south, h can be calculated by applying the sine rule to the small spherical triangle (centre of sphere at P) Fig. 7.10.

$$\frac{\sin \sigma}{\sin h} = \frac{\sin(90-\phi)}{\sin 90} = \frac{\cos \phi}{1}$$

$$\sin h = \frac{\sin \sigma}{\cos \phi} = \frac{\sin 16.87}{\cos 51}$$

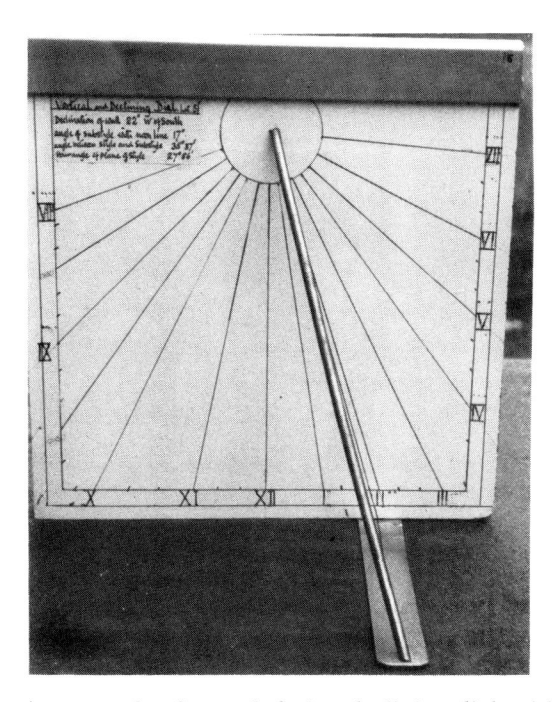

$$h = 27°.46.$$

Fig. 7.11. An example of a south facing declining dial, with a copper rod style, held in position by a support, perpendicular to the plane of the dial. The shadow angles are those given in Table 7.3. Declining angle 22° west of south, latitude 51°.

The south facing dial formula for calculating shadow angle γ, is from Section 7.7

$$\tan \gamma = \tan H \cos \phi.$$

For a dial declining an angle 22° west of south, H is modified to $(27.46-H)$ and the tangents of the shadow angles are reduced by the factor $\cos \theta = \cos 22°$. The calculation of the shadow angles can now be made.

$(h-H)$ can be found for each hour angle (taking H as negative in the forenoon) from,

$$\tan \gamma = \tan(27.46-H) \cos 51° \cos 22° \qquad [7.9(3)]$$

(for ease of calculation on the calculator, $\cos 51° \cos 22° = 0.5835$ this can be stored in the 'memory' of the calculator).

Table 7.3. The shadow angles are measured from the substyle line which is itself 16°.87 to the east of the vertical. The negative sign indicates an anti-clockwise measurement.

Sun time	H Hour angle of Sun	$(h-H)$ $(27.47-H)$	Shadow angle γ of Sun measured from the substyle line from [7.9(3)]
0800	−60	87.47	85.67
0830	−52.5	79.97	73.14
0900	−45	72.47	61.57
1000	−30	57.47	42.45
1100	−15	42.47	28.11
1200	0	27.47	16.87
1300	15	12.47	7.35
1400	30	−2.53	−1.48
1500	45	−17.53	−10.44
1600	60	−32.53	−20.41
1700	75	−47.53	−32.51
1800	90	−62.53	−48.30
1900	105	−77.53	−69.24

7.11 Altitude Sundials

There are three common types of sundial that depend essentially on the Sun's altitude. These dials are of great antiquity as early astronomers were mainly concerned with measuring the altitudes of heavenly bodies. The dials can now be drawn very easily from a table of figures that can be compiled by the use of a calculator (see Fig. 7.11). As these sundials

Projects with Sundials and the Calculator

depend only on the Sun's altitude, a table can be drawn up using the basic spherical trig formula (Table 7.4):

$$\sin(\text{alt}) = \sin\phi \sin\delta + \cos\phi \cos\delta \cos(\text{HA}) \qquad [7.11(1)]$$

We use for convenience of design the declinations of the Sun corresponding to the approximately equally spaced monthly intervals of time for the year from 21st June, as shown in Table 7.4.

The pillar, the disc, and the Capuchin sundials, all depend on the *Sun's altitude* and the relation [7.11(1)], or the rearrangement of it:

$$\cos(\text{HA}) = \frac{\sin(\text{alt}) - \sin\phi \sin\delta}{\cos\phi \cos\delta}.$$

Telling the time by measuring the Sun's altitude was specially prevalent in latitudes within about 30° or 40° of the equator and even today in rural parts in tropical regions people express a time of day by holding up an arm to indicate the Sun's altitude.

In northern latitudes people used the Sun's direction or azimuth as a useful way of indicating roughly the time.

In order to measure the altitude of the Sun or heavenly body a quarter of

Table 7.4. Pillar dial coordinates for curves (corresponding altitudes of the Sun in brackets) Lat. 51°.

Date	Sun's declination in degrees	(Noon) 00	(11–13) 15	(10–14) 30	(9–15) 45	(8–16) 60	(7–17) 75	(6–18) 90	(7–19) 105
	x axis			y ordinates = 85 tan(alt)					
21st Jun	23.44	162.94 (62.45)	147.78 (60.1)	117.05 (54.015)	87.53 (45.84)	63.39 (36.71)	43.86 (27.29)	27.63 (18.00)	13.75 (9.19)
21st Jul 24th May	20.6	144.88 (59.60)	132.9 (57.40)	107.20 (51.59)	81.02 (43.62)	58.66 (34.49)	40.01 (25.21)	24.16 (15.86)	10.36 (6.95)
22nd Aug 21st Apr	11.83	104.33 (50.82)	97.65 (48.96)	81.65 (43.85)	62.90 (36.50)	44.92 (27.85)	28.58 (18.58)	13.72 (9.17)	0
23rd Sept 21st Mar	0	68.83 (39.0)	65.07 (37.44)	55.25 (33.02)	42.24 (26.42)	28.18 (18.34)	14.03 (9.37)	0	
24th Oct 18th Feb	−11.7	43.87 (27.29)	41.37 (25.95)	34.5 (22.09)	24.61 (16.15)	12.92 (8.64)	2.71 (1.826)		
22nd Nov 22nd Jan	−19.9	29.43 (19.09)	27.42 (17.87)	21.75 (14.35)	13.24 (8.85)	2.67 (1.80)	−9.52 —		
21st Dec	−23.44	23.65 (15.54)	21.81 (13.94)	16.52 (11.00)	8.47 (5.7)	−1.73 —			

a circle graduated in degrees is adequate, as we require only the arc from the horizon to the zenith (see Fig. 3.5).

The French word for a sundial is *Cadran* (or *Cadran Solaire*) and the word *Cadran* comes from 'quadrant' meaning a quarter of a circle, which was an early means of measuring altitudes (see Section 5.11).

7.12 The Pillar Dial

The making of a pillar dial is a useful and instructive project (see Fig. 7.12(a) and Table 7.4).

The altitudes corresponding to various declinations of the Sun from −23°.44 to 23°.44 are calculated for a particular latitude (in Fig. 7.12(a) this is 51°N). The table of results thus obtained is shown in Table 7.4 and these are plotted on a sheet of centimetric graph paper, size A4 (see Fig. 7.13). The paper is then wrapped round a cylinder of wood or a suitable canister having a diameter of just over 80 mm. A rotatable arm shown in Fig. 7.12(a) forms a style by projecting over the edge of the cylinder 85 mm, so that,

$$\tan(\text{altitude of the Sun}) = \frac{\text{length of shadow}}{85}.$$

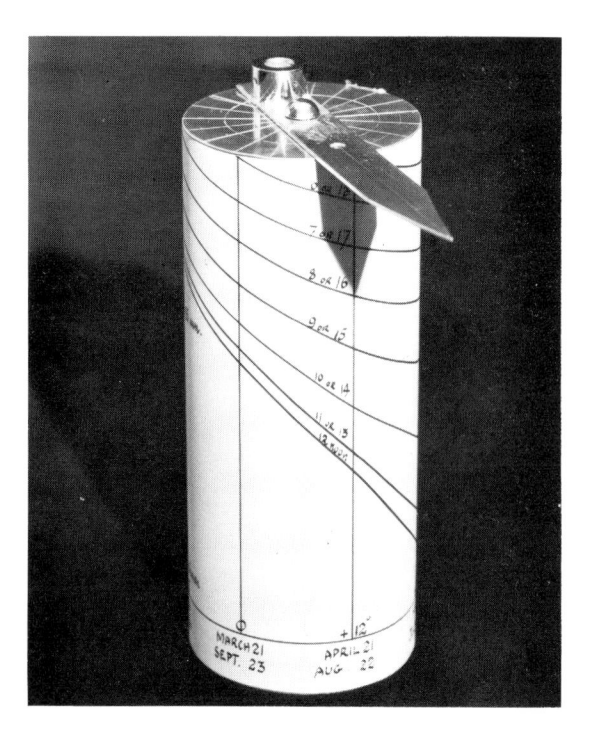

Fig. 7.12(a)

Projects with Sundials and the Calculator

The style is rotated on the pillar until it points in the direction of the date inscribed round the bottom of the pillar. With the style still on the date, turn the pillar until the pointed style points directly in the direction of the Sun, in azimuth, the pillar remaining upright.

A table of values calculated for making a dial for lat 51° is given in Table 7.4.

This type of dial is of ancient origin and a fine example of 17th century craftsmanship is shown in Fig. 7.12(b), by courtesy of the Science Museum, London.

Fig. 7.12(b)

Crown Copyright: Science Museum, London

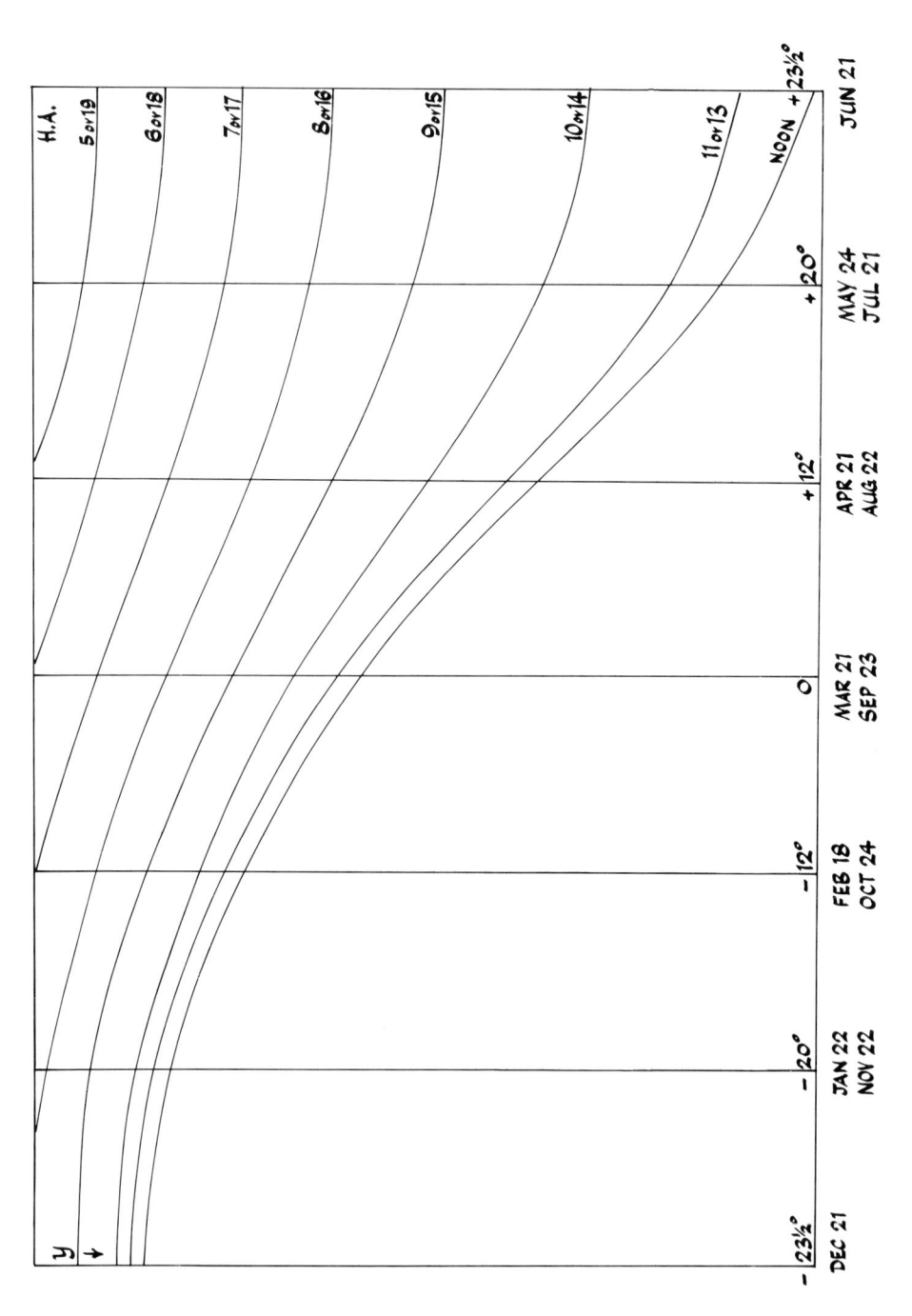

Positional Astronomy and Astro-Navigation Made Easy

Fig. 7.13. Curves for pillar dial.

Projects with Sundials and the Calculator

Table 7.5. Co-ordinates for a disc dial.

Dates	Declination	Value of x on graph	(Altitude at 12^h) y ordinate $= (85 + \frac{N}{1.5})\tan(\text{alt})$ HA = 0	\multicolumn{8}{c}{Sun Time and Hour angle (altitudes are in brackets)}								
				11 or 13 (15°)	10 or 14 (30°)	9 or 15 (45°)	8 or 16 (60°)	7 or 17 (75°)	6 or 18 (90°)	5 or 19 (105°)
21 June	23.44	85	(62.44) 162.87	(60.1) 147.81	(54.02) 117.08	(45.84) 87.53	(36.71) 63.38	(27.29) 43.85	(18.00) 27.62	(9.19) 13.75
21 July 24 May	20.6	105	(59.60) 178.96	(57.4) 164.18	(51.59) 132.42	(43.216) 100.06	(34.49) 72.13	(25.21) 49.43	(15.86) 29.83	(6.95) 12.80
22 Aug 21 Apr	11.83	125	(50.82) 153.37	(48.96) 143.59	(43.85) 120.08	(36.50) 92.50	(27.85) 66.04	(18.58) 42.02	(9.17) 20.18	
21 Sept 21 Mar	0	145	(39) 117.4	(37.44) 110.86	(33.02) 94.23	(26.42) 72.04	(18.34) 48.07	(9.37) 23.93	0	
24 Oct 18 Feb	−11.7	165	(27.29) 85.13	(25.95) 80.30	(22.09) 66.97	(16.15) 47.78	(8.64) 25.07	(1.826) 5.260		
22 Nov 22 Jan	−19.9	185	(19.09) 64.03	(17.87) 59.65	(14.35) 47.32	(8.85) 28.80	(1.80) 5.813			
21 Dec	−23.44	205	(15.54) 57.00	(13.94) 50.88	(11.00) 39.84	(5.7) 20.46				

185

Positional Astronomy and Astro-Navigation Made Easy

7.13 The Disc Dial

The disc dial is easy to make and is another simple device for measuring the altitude of the Sun, and thereby calculating the time.

In Fig. 7.14, A is a pin at the top left hand corner of a square cut from a piece of hardboard 200 mm square. In operation the square disc is held so that the plane of the disc is vertical and in the plane of the Sun from the observer. The shadow of the pin made by the Sun thus falls across the disc as shown in the figure. The top edge of the square AD is held horizontally using a spirit level along AD, or a plumbline to keep the side AC vertical. The dial curves are plotted from the results calculated and shown in Table 7.5.

A convenient dial can be made by plotting the points on centimetre graph paper with the number of days between 21st December and 21st June along the x-axis and the x tan (altitude of the Sun) along the y-axis.

To use the device, hold steadily as above and note where the shadow of the pin intersects the y-axis corresponding to the date. The time is given by this point, by noting where it lies with respect to the hour angle curved lines.

For example, on 21st July and on 24th May the pin's shadow cuts this date ordinate on the 8 HA—or 16 HA—curve, thus giving the time.

The table of coordinates for the hour angle curves was drawn up as follows.

A convenient scale is as shown in Fig. 7.14 with the ordinate or y-axis for 21st June being drawn 85 mm from the vertical axis AC. The months are spaced at intervals of 20 mm as shown.

Then

$$x = \left(85 + \frac{N}{1.5}\right)$$

where N is the number of days since 21st June (30 days = 20 mm).

For example the coordinates of the points lying on the curve for (HA) = 45°, i.e., for 0900 and 1500, are calculated from the Sun's altitudes at this HA at the various times of the year, or what amounts to the same thing, at various declinations of the Sun.

Consider the latitude 52°N, (HA) = 45° on 24th May, i.e., (δ) = 20°.6N. The Sun's altitude by our familiar formula [3.7(1)] is

$$\sin(\text{alt}) = \sin 52 \sin 20.6 + \cos 52 \cos 20.6 \cos 45$$
$$(\text{alt}) = 43°.216$$

Projects with Sundials and the Calculator

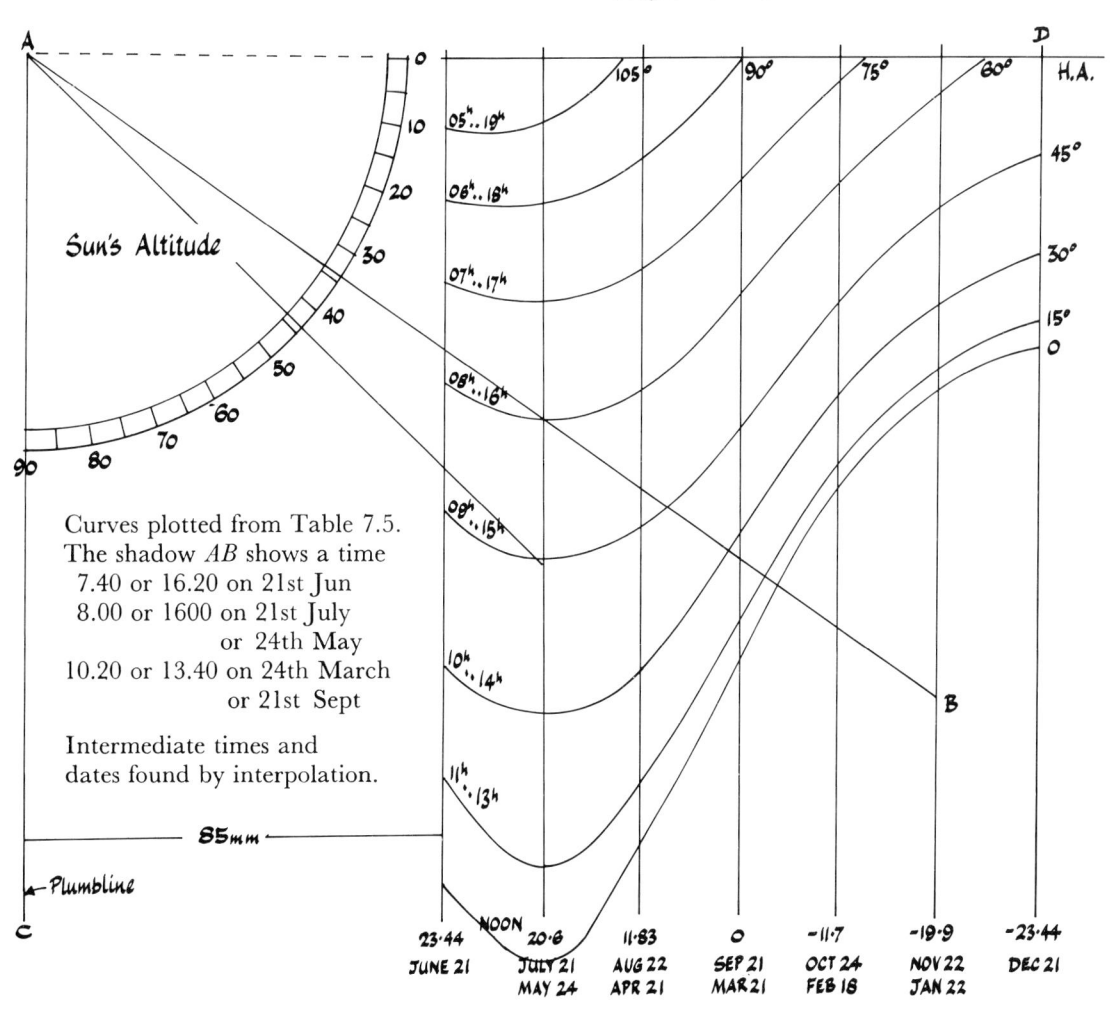

Fig. 7.14

but, $y/x = \tan 43.216$
         $x = 85 + 20 = 105$
         $y = x \tan 43.216$
therefore $y = 98.8$ mm.

To complete the (HA) = 45° curve, find the altitudes for declinations corresponding to the dates on the graph. These altitudes can be calculated as above. Thus we can obtain the y ordinate from

$$y = \left(85 \text{ mm} + \frac{N}{1.5}\right)\tan(\text{alt})$$

where N is the number of days elapsed since 21st June and reckoning that

187

Positional Astronomy and Astro-Navigation Made Easy

30 days occupy 20 mm on the graph. Tabulated results for the drawings of the hour angle curves are given in Table 7.5.

7.14 It will be noticed that the graphs for the disc dial and the pillar dial deal with the same information, and translate altitudes of the Sun into corresponding hour angles or sun times.

This suggests a project for drawing a simple set of curves shown in Fig. 7.15, showing the relation between the sun time and the altitude in *degrees* which we can easily measure with one of the devices in Section 3.2, or by the stick in the sand device of Section 7.2.

The graphs and a protractor can be used as a reasonably accurate means of telling the time, i.e., as a light, portable altitude sundial. For example, you observe the Sun's altitude is 25° in the middle of May. The time is indicated by the small cross and falls a little above the 7 or 17 hour angle curve. By interpolating between the curves the time is 7.10 a.m. or 4.50 p.m. It is assumed that you will know whether it is morning or evening!

7.15 The De Saint Regaud or Capuchin Dial

It is of interest to note here a disc type *altitude* sundial by De Saint Regaud which had its origin in the 15th Century and is often called the Capuchin dial. It reveals the ingenious geometry used to overcome the difficulty of the spherical triangle. This dial is described more fully in the book *Sundials* by Frank Cousins (see Bibliography).

The Sun's altitude is measured by means of a sighting bar and a plumbline, and the geometry of the marking out uses the latitude of the place and the Sun's declination.

The construction of the dial markings in effect solves the basic spherical triangle [3.7(1)]

$$\cos(\text{HA}) = \frac{\sin(\text{alt}) - \sin\phi \sin\delta}{\cos\phi \cos\delta}$$

by a geometrical slide rule–calculator device. In Fig. 7.16 the angles, altitude, ϕ, δ and HA are marked.

From the construction, OPL is the plumbline and $OP = OA$. XPB is a perpendicular on AC.

The angle ACX is the LHA of the Sun and the point where BX cuts the arc AE at T marks the time. In Fig. 7.16 this is $\approx$ 4.15 p.m. or 7.45 a.m.

$$\cos(\text{LHA}) = \frac{PN}{CX} \quad \text{but} \quad PN = PQ - NQ$$
$$= \frac{PQ - NQ}{CX}$$

188

Projects with Sundials and the Calculator

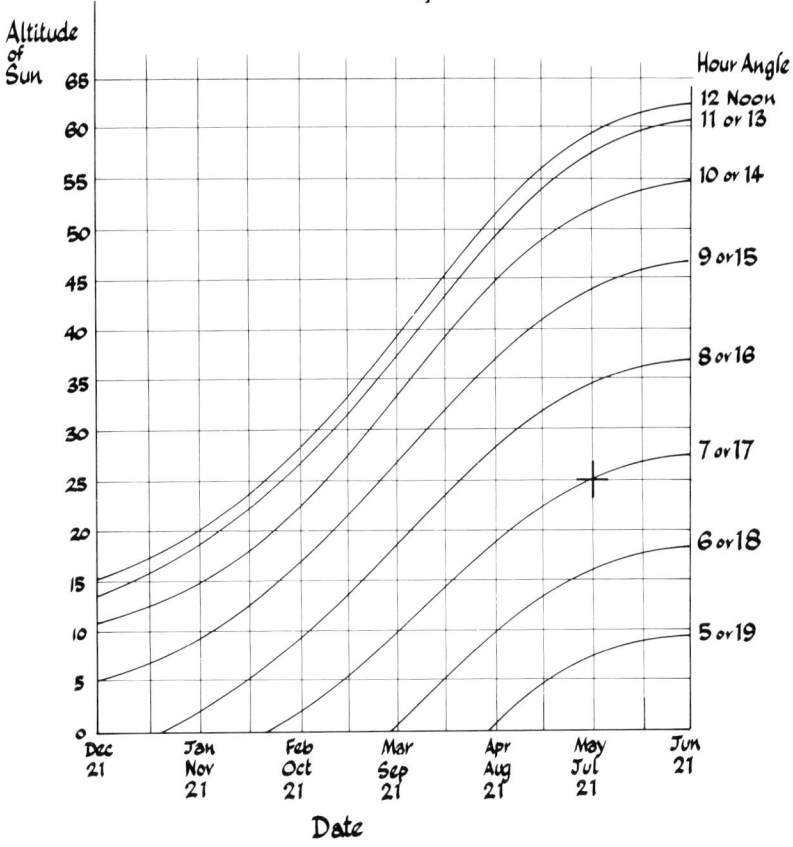

Fig. 7.15. Graphs showing how the altitude of the Sun changes with the time of the day (Sun's hour angle) at various times of the year. Drawn for latitude 51°.

but,
$$PQ = OP \sin \alpha$$
$$NQ = OF = DO \sin \phi$$
$$DO = OA \sin \delta$$
$$NQ = OA \sin \delta \sin \phi$$

also,
$$CX = CA = DA \cos \phi$$
$$DA = OA \cos \delta$$

$\therefore$
$$CX = OA \cos \alpha \, \cos \phi$$

$\therefore$
$$\cos(\text{LHA}) = \frac{OA\,(\sin \alpha - \sin \delta \sin \phi)}{OA(\cos \alpha \, \cos \phi)}$$
$$= \frac{\sin(\text{alt}) - \sin \phi \sin \delta}{\cos \text{alt} \cos \phi}.$$

which is the relation [3.7(1)].

189

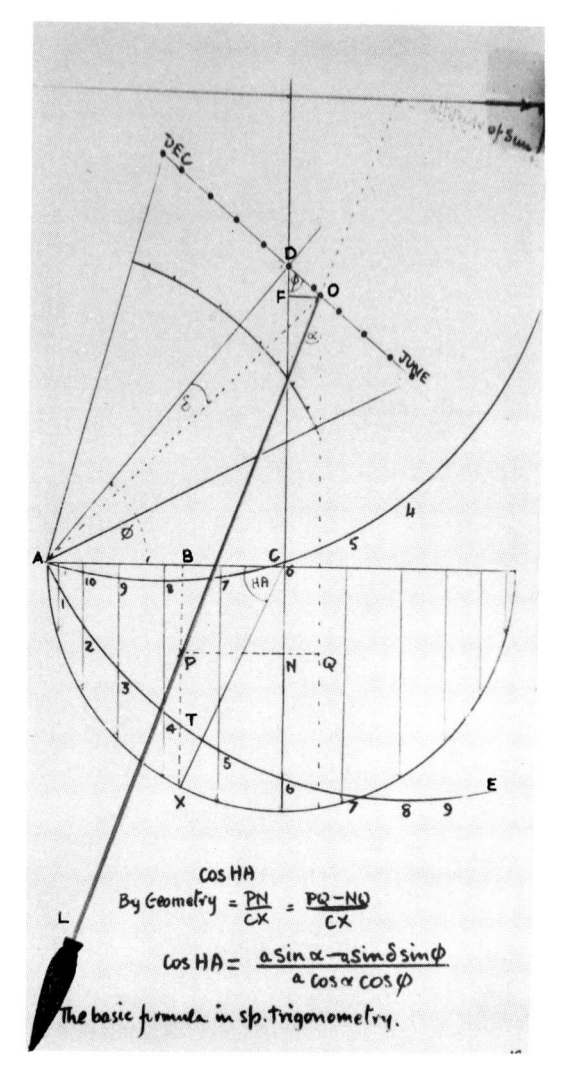

Positional Astronomy and Astro-Navigation Made Easy

Fig. 7.16

The hand-held electronic calculator takes the input of altitude, ϕ and δ and displays the hour angle (sun time) in a few seconds, but the graphical solution of the spherical trigonometrical relation is fascinating.

7.16 The Polar Dial

The style is fixed perpendicularly to the dial, as shown in Fig. 7.17, and points to the equator, since it is at right angles to the polar axis. In Fig. 7.17 the length of the style S is 50.8 mm.

The polar dial depends on the Sun's hour angle and uses the following

Projects with Sundials and the Calculator

relations for the x and y coordinate of the shadow tip, given by,

$$x = S \tan H \qquad y = S \frac{\tan \delta}{\cos H}.$$

These relations can best be established by a simple model and using thread and 'Blu-Tac' or Plasticine to represent directions of the Sun and shadow angles. The curves for each declination are readily calculated using a calculator. The tabulated results are given in Table 7.6.

The hour lines for various declinations are independent of the latitude since in all latitudes the dial is in the plane of the polar axis and lies east–west. The noon line of the dial points to the celestial pole.

In Fig. 7.18, $\delta =$ declination of the Sun, $x = 50.8 \tan H$ and $y = CA \tan \delta$

but
$$CA = \frac{50.8}{\cos H}$$

$$y = 50.8 \frac{\tan \delta}{\cos H}.$$

Table 7.6. Polar sundial

Hour angle H Time		Degrees	Values of x 50.8 tan(HA) mm	y for δ ±23.44	y for δ ±20.2	y for δ ±12	δ 0
12 noon		0	0.00	22.03	18.69	10.80	0
11.30	12.30	7.5	6.688	22.22	18.85	10.89	0
11.00	13.00	15.00	13.612	22.80	19.35	11.18	0
10.30	13.30	22.5	21.04	23.84	20.23	11.68	0
10.00	14.00	30.00	29.32	25.43	21.58	12.47	0
09.30	14.30	37.5	38.980	27.76	23.56	13.61	0
09.00	15.00	45.00	50.80	31.15	26.43	15.27	0
08.30	15.30	52.5	65.02	36.25	30.70	17.74	0
08.00	16.00	60.00	87.988	44.05	37.38	21.60	0
07.45	16.15	63.75	103.01	49.80	42.26	24.41	0
07.30	16.30	67.15	122.64	57.55	48.8	28.21	0
07.15	16.45	71.25	149.65	68.34	58.15	33.59	0
07.00	17.00	75.00	189.58	85.1	72.21	41.72	0
06.45	17.15	78.75	255.386	112.9	95.80	55.35	0

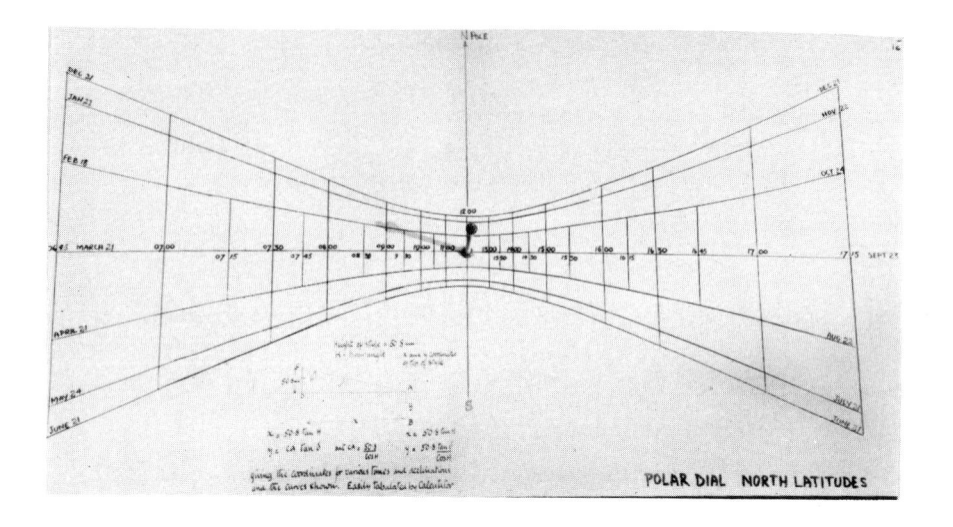

Positional Astronomy and Astro-Navigation Made Easy

Fig. 7.17

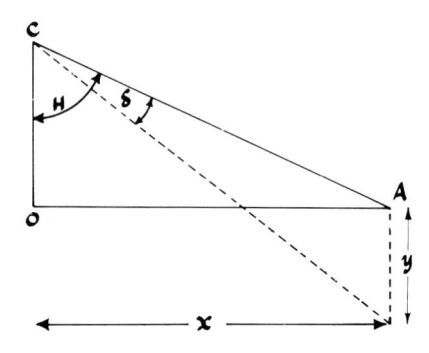

Fig. 7.18

7.17 An Azimuth Dial

The Elliptical or Analemmatic Dial (horizontal dial with a vertical style).

The construction and marking of this dial provides a further project involving the calculator which facilitates the making of tables of angles, and so eliminates the usual complicated geometrical constructions associated with this dial.

We imagine a circular dial in the equatorial plane to be projected on to a horizontal plane. This produces an ellipse with semi-major axis a and minor axis b, so that

$$b = a \sin \phi \quad \text{(where } \phi = \text{the latitude)}.$$

It is therefore easy to construct this ellipse on graph paper so that all

Projects with Sundials and the Calculator

points on the graph satisfy the relation
$$x = y \sin \phi \quad \text{(see Fig. 7.19)}.$$

The analemmatic dial is essentially an *azimuth* device so we apply the relation [3.7(8)] which does not involve altitudes:

$$\cot(\mathrm{Az}) = \frac{\tan \delta \cos \phi - \sin \phi \cos(\mathrm{HA})}{\sin(\mathrm{HA})}.$$

When $\delta = 0$ the vertical style is at the centre of the ellipse and the shadow angles (Az) mark the hours on the periphery of the circle (the equator) of which the ellipse is the projection.

Then
$$\cot(\mathrm{Az}) = -\sin \phi \cos(\mathrm{HA})$$

or
$$\tan(\text{shadow angle}) = \frac{\tan(\mathrm{HA})}{\sin \phi}. \qquad [7.17(1)]$$

We can thus mark the hour lines on the ellipse and these are correct for the vertical style placed at the centre, but for the declination 0° only. To make the dial usable for other declinations we have to move the style along AB.

The position of the moveable style on the line AB so as to give the correct shadow times depends on the declination and it can be shown that the required distance from the centre is D

$$= a \tan \delta \cos \phi \qquad [7.17(2)]$$

where a is the semi-major axis of the ellipse.

Books on sundials often avoid any attempt to establish this formula although it follows very simply from the principles on which the dial is constructed.

Let V be the position of the vertical style which gives the correct shadow angle for the corresponding hour angle. Then, from Fig. 7.19, the distance

$$D = OV = UO - UV$$
$$= RQ - UV$$

But, $PQ = a \cos(H)$. Also, $RQ = PQ \sin \phi$. ((H) is the hour angle)

∴ $\qquad RQ = a \cos(H) \sin \phi$

So, $\qquad OV = a \cos(H) \sin \phi - UV \qquad [7.17(3)]$

Also, $\qquad UV = OQ \cot \gamma \quad$ and $\quad OQ = a \sin(H)$.

The shadow angle, γ, which is also the Sun's azimuth, in this type of

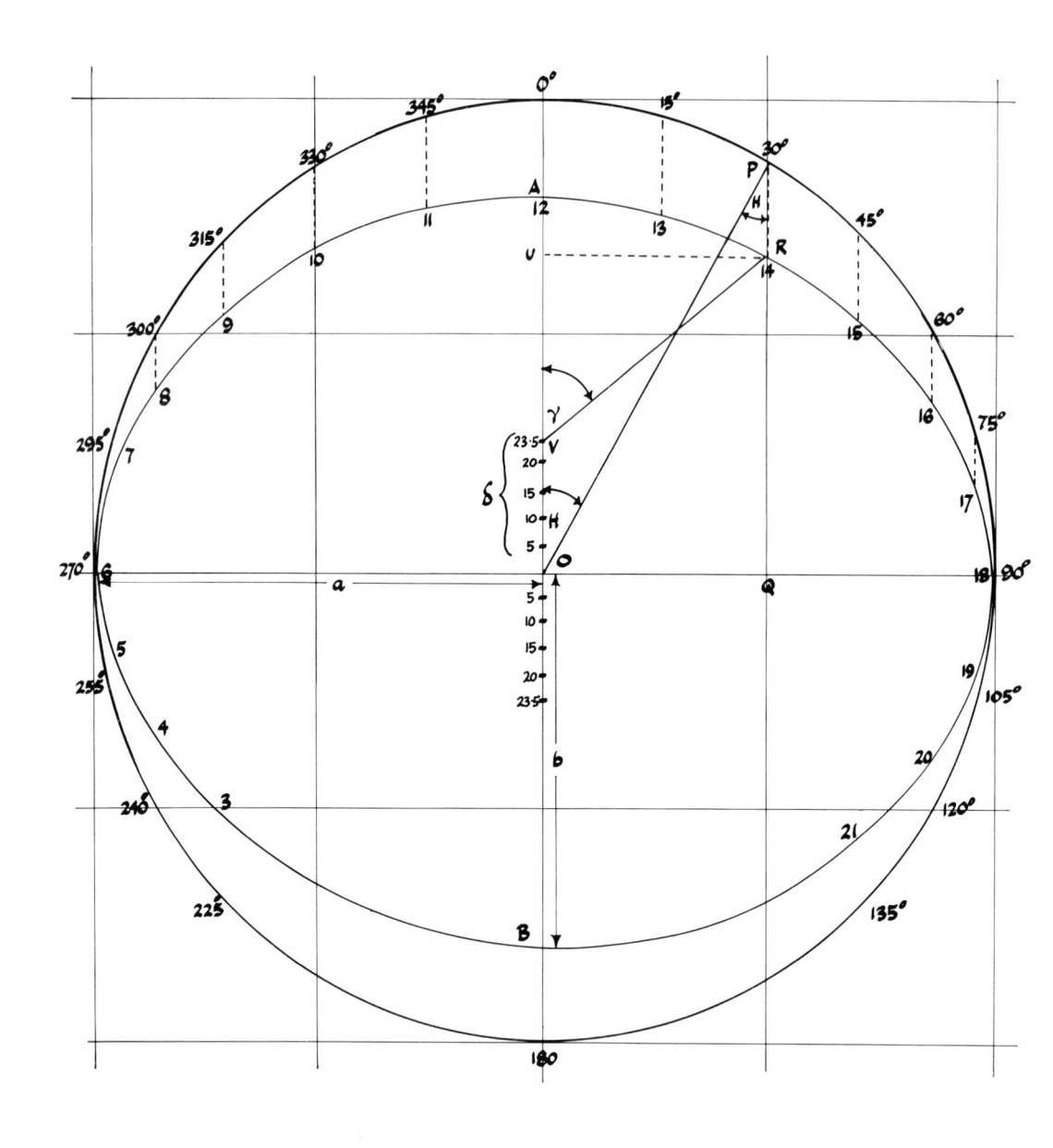

Fig. 7.19. Analemmatic dial. Lat 51° N or S

$$\frac{b}{a} = \sin \phi \qquad \tan \gamma = \frac{\tan h}{\sin \phi} \qquad D = a \tan \delta \cos \phi$$

Projects with Sundials and the Calculator

dial is given by the relation 3.7(8)
$$\cot \gamma = \frac{\cos \phi \tan \delta - \sin \phi \cos(\text{HA})}{\sin(\text{HA})}$$
therefore from (3) above,
$$OV = a \cos(\text{H}) \sin \phi - OQ \cot \gamma$$
$$= a \cos(\text{HA}) \sin \phi - a \sin(\text{H}) \left(\frac{\cos \phi \tan \delta - \sin \phi \cos(\text{HA})}{\sin(\text{HA})} \right) \text{ this readily}$$
simplifies to give:

distance of style D from $O = a \cos \phi \tan \delta$

Tables (1) and (2) of 7.18 contain all the information necessary for the construction of this dial. Figure 7.19 shows the analemmatic dial and the positions of the vertical style for different times of the year, corresponding to the Sun's declinations.

7.18 For an analemmatic dial having a semi-major axis of 100 mm and designed for latitude 51° we calculate:

(1) The position of the shadow angles from the centre from:
$$\tan(\text{shadow angle}) = \frac{\tan(\text{HA})}{\sin \phi}.$$

Sun time HA	Shadow angle
0°	0°
15	19.02
30	36.61
45	52.15
60	65.83
75	78.24
90	90.00
105	−78.24

(2) Position of the style for various declinations from centre
$$= 100 \tan \delta \cos \phi.$$

Declinations	Distance of style from C
± 0	0
± 5	± 5.5 mm
±10	±11.1 mm
±15	±16.86 mm
±20	±22.9 mm
±23.44	±27.36 mm

These tables give only the main values required, but the calculator can provide all intermediate values for an accurate marking for this fascinating dial.

7.19 A Suggested Project Concerning the Analemmatic Dial, which may Provoke Discussion, and will Call for Accurate Observations

It is stated in many authoritative works on sundials, e.g., the *Encyclopaedia Britannica*, that this dial can be used in conjunction with an ordinary horizontal dial for finding both the time and the meridian without a compass. It is generally claimed that the double instrument is able to be correctly orientated with respect to the north–south meridian simply by turning the plate in the horizontal plane until the analemmatic and the horizontal dial each read the same hour (Fig. 7.20).

It is of course clear that both dials give the correct time when their meridian lines are correctly orientated N–S—they are designed to do so—but it is claimed that they will record 'different times in any other position'. This is only partially true and this note refers to the limits beyond which it is unsatisfactory to accept this claim.

The results of a study of what happens to the reading on a horizontal dial

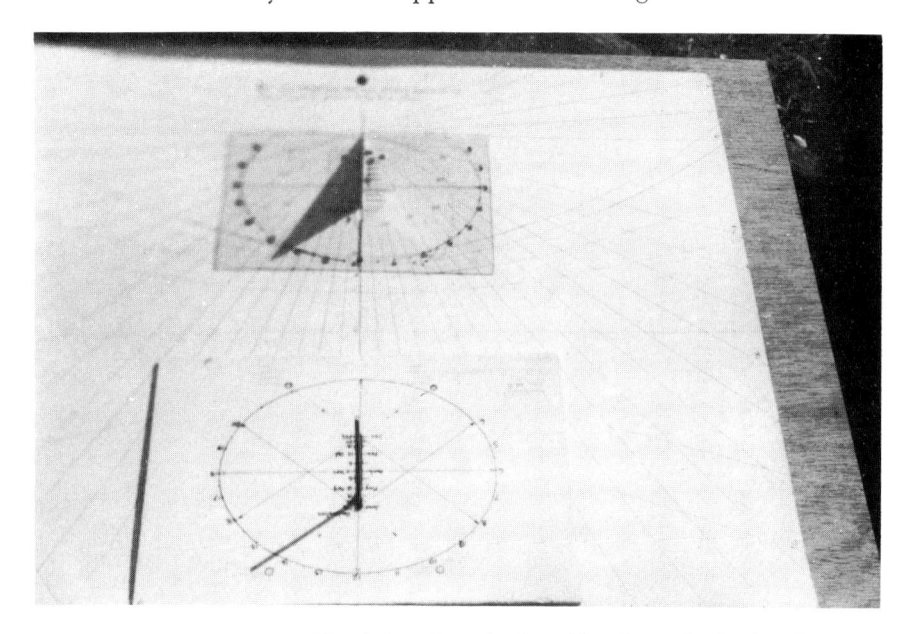

Fig. 7.20. Practical verification of calculated results.

Projects with Sundials and the Calculator

when rotated in a horizontal plane by say 15° has been published recently and these results showed a wide range, depending on the HA and the declination. This prompted the enquiry into what really happens when an analemmatic dial (combined with its companion horizontal dial) is searching for equal readings and therefore the true position of the meridian. As the analemmatic dial is an azimuth instrument, the effect of giving it a twist of 15° about its vertical style can easily be calculated, as the twist merely changes the Sun's apparent azimuth, as seen by the sundial.

Between the hours of 11.00 a.m. to 12.00 p.m. it is impossible to get any satisfaction from the combined instrument for nearly equal times will be recorded over a wide range of positions (15° on either side of the meridian) as the graphs will show. Times recorded differ by only two or three minutes from 10.00 a.m. to 12.30 p.m., and sundials themselves are accurate only to within two or three minutes because of the diameter of the Sun and diffraction effects that accompany shadows. Therefore a search for the time on both instruments leaves the observer with a range of positions to choose both the time and/or the orientation.

From before 10 a.m. and after 1 p.m. the combined instrument functions as claimed, and reaches optimum conditions with HA's of 60° or 300°, i.e., 1500^h or 0900^h.

The graphs show the way in which each part of the instrument behaves when it is 15° off meridian.

(1) Working at $\delta+20$.
(2) at $\delta-0$.
(3) δ at -20.

The results shown in tabulated form were calculated from standard spherical trig relations and checked both graphically and by a simulated Sun model.

The analemmatic results were calculated from relation [3.7(8)]

$$\cot(Az) = \frac{\tan \delta \, \cos \phi - \sin \phi \, \cos(HA_1)}{\sin(HA_1)}$$

using HA's ranging from 0 to 105. Having found this azimuth (Az) then add 15° to it and put the new Az back into the relation and calculate the new HA, HA_2.

The difference between HA_2 and HA_1 is the advancing effect of the twist to 15° off meridian. These values are recorded in Table 7.7 and are shown in graphical form in Fig. 7.21. They indicate that between the hours of 10 and 13 it would be very difficult to make any practical use of the fact, that the two dials record different times, *unless* both dials in combination are correctly orientated. The trouble is they can record the same times when *not* correctly orientated, and round about the middle of the day the error can be as dangerous as that displayed in the unskilled use of the 'watch' compass of Section 5.14.

For example, at 10 a.m. while both dials are 15° out of proper orientation, the dials show only about 4 minutes difference in their times, at 11 a.m. and at noon it would be impossible to detect any significant difference in times, at 1 p.m. the difference becomes about 5 minutes which could just be detected, although both the dials are 15° off the meridian. So from 10 a.m. to 1 p.m. the two dials give approximately the same time although the dials are both 15° off the meridian. Thus one could be nearly 3 hours out in time and 15° out in compass bearing.

The combination of dials do however give useful results for all declinations in the early morning, i.e., from sunrise to 10 a.m. and from 1 p.m. to sunset.

Table of results obtained by calculation and confirmed by a practical experiment and which are represented graphically in the three graphs, Fig. 7.21.

When a horizontal dial is given an anti-clockwise twist about a vertical axis (contrary to all sundial rules!) the time on the sundial is advanced by the number of minutes shown in columns under each of the three declinations, 20°, 0, and −20°. The corresponding times of advance on the analemmatic dial are shown in adjoining columns.

Table 7.7. Minutes of advance in sundial readings caused by a 15° twist of a combined horizontal–analemmatic dial.

Local sun time	Declination +20 Analemmatic dial	Declination +20 Horizontal dial	Declination 0° Analemmatic dial	Declination 0° Horizontal dial	Declination −20 Analemmatic dial	Declination −20 Horizontal dial
6	75	45.4	75	49		
7	70	44	70	48.5		
8	61	39.3	68.4	48	70	58.3
9	50	38	56.8	47.2	67	59.8
10	39	36.6	49	46.6	65	60.6
11	35	34.3	46	46.2	60	60.2
12	33.6	32.8	45	46.4	60	60
13	38	31.8	50	46.4	65	58
14	52	32.8	59	46.4	68	56
15	65	36	65	46	73	53
16	80	40.6	72	45.8	80	49.3
17	85	43	75	45.5		
18	85	48.7				

Minutes advanced in each dial by a 15° anticlockwise rotation from Meridian

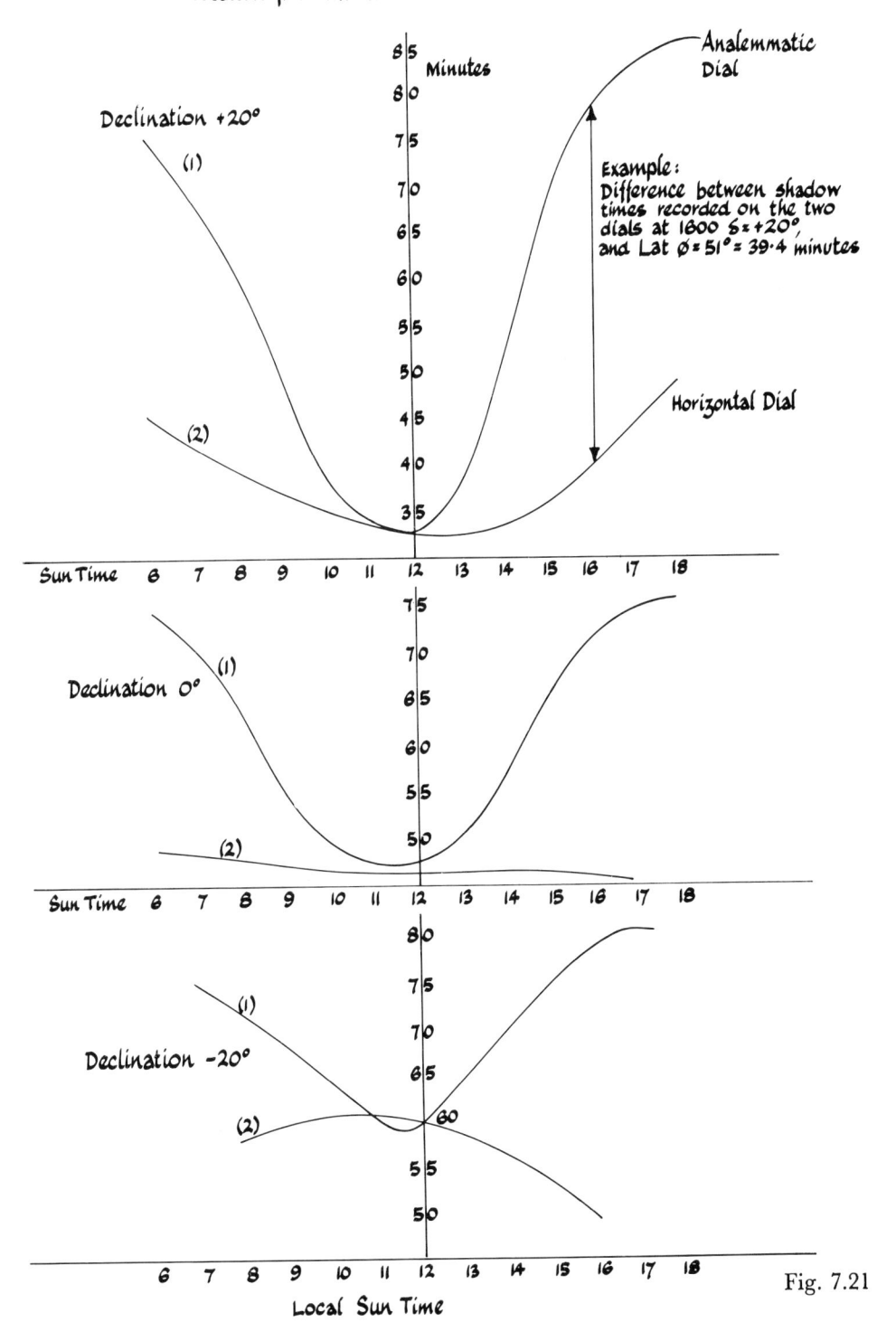

Fig. 7.21

Positional Astronomy and Astro-Navigation Made Easy

The graphs show that the claim that the combination of a horizontal dial and an analemmatic dial will serve as a compass is not well founded when the hour angle of the Sun is small, i.e., between 10 a.m. and 1 p.m.

7.20 Declination or Zodiacal Lines

If a small bead called a nodus is stuck on the tip of a style of a vertical or horizontal sundial, the position of the shadow of the nodus on the dial will not only indicate the time, according to the hour angle line it falls on, but it can also indicate the approximate date, or the Sun's declination, which varies with the date. The bead's shadow during its daily passage across the dial will travel along a line of declination or a zodiac line. The Sun's declination is associated with the Sun's position in the ecliptic or zodiac.

Complicated geometrical constructions are given for these lines, but the drawing of the lines can be easily done by simply plotting the lines from the rectangular coordinates, x and y, using the top tip of the style, A, as origin. AF is the y-axis.

The calculation of the coordinates and plotting the points could be a project and an exercise (see Fig. 7.22).

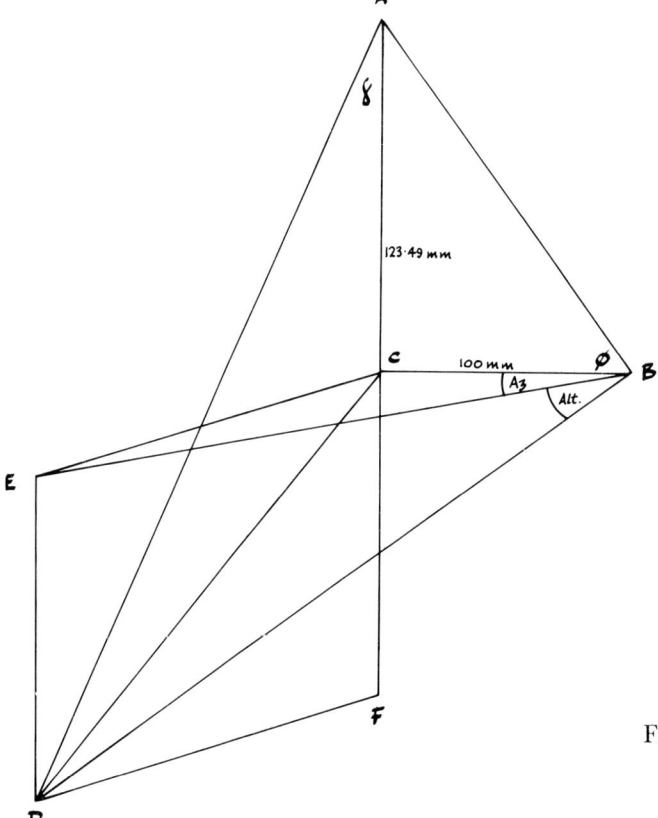

Fig. 7.22

Projects with Sundials and the Calculator

A table of coordinates for the zodiacal lines of a vertical south facing dial has been compiled for the hour angles 0, 15, 30 and 45. After HA 45° the lengths of the y ordinates become increasingly large (see Table 7.8).

Fig. 7.22 represents a vertical south facing dial and AB is the style. Seven declination lines for this dial are shown in Fig. 7.23 corresponding to declinations $-23°.44$, $-20°$, $-10°$, $0°$, $10°$, $20°$ and $23°.44$. The co-ordinates for plotting these curves can be calculated from the relations derived from the figure (see Table 7.8).

AB is the style of a vertical south facing declining dial. B is the nodus, and D is the shadow point of B. $ACFDE$ is the plane of the dial. CB is 100 mm.

CB is arbitrarily taken in the results tabulated as 100 mm, but the scale can be changed to suit the size of paper or wall space available.

AF is the y ordinate of D, A is the origin, FD is the x abscissa of D and γ is the shadow angle of the style.

Then $FD/AF = \tan \gamma$ but $FD = CE = CB \tan(\text{Az}) = x$

Therefore $\qquad y = AF = \dfrac{FD}{\tan \gamma} = \dfrac{CB \tan(\text{Az})}{\tan \gamma}$

but $\tan \gamma = \tan(\text{HA}) \cos \phi$ ... from the relation for a vertical south facing dial (see Section 7.7).

The y ordinate is therefore given by

$$y = \dfrac{CB \tan(\text{Az})}{\tan(\text{HA}) \cos \phi} \quad \text{or} \quad \dfrac{100 \tan(\text{Az})}{\tan(\text{HA}) \cos \phi}$$

and $\qquad x = 100 \tan(\text{Az})$.

It is interesting, and a useful check on accuracy, to find out just where the zodiacal lines cut the horizontal line through B. This takes place when the shadow of the nodus B is horizontal, i.e., when the altitude of the Sun is 0, that is, when it is rising or setting. For an east–west horizontal dial this can only take place when declinations are less than 0, i.e., for negative declinations. Then,

$$\cos(\text{Az}) = \dfrac{\sin \delta}{\cos \phi} \quad \text{(from relation [3.7(6)])}.$$

If, for example $\delta = -20$, then

$$\cos(\text{Az}) = \dfrac{\sin -20°}{\cos 51}$$

$$(\text{Az}) = 122°.92.$$

Then the x coordinate for the last appearance of the shadow of the nodus will be at $x = 100 \tan 122°.92 = 154.458$ mm from C.

Positional Astronomy and Astro-Navigation Made Easy

Table 7.8. Table for plotting the declination lines or zodiacal lines on a vertical south facing dial. Nodus 100 mm from dial face. The coordinates x and y are found as follows: for $(HA) = 0$, $x = 0$ and $y = 100\tan(alt)$. For other hour angles, $x = 100\tan(Az)$, $y = \dfrac{x}{\tan(HA)\cos\phi}$ or $\dfrac{100\tan(Az)}{\tan(HA)\cos\phi}$. $\phi = 51°$.

Declination	(HA) = 0			(HA) = 15° or 345°			(HA) = 30° or 330°			(HA) = 45° or 315°		
	x	Meridian transit altitude	$y = 100\tan(alt)$	Az of shadow	x	y	Az	x	y	Az	x	y
−23.44	0	15.56°	27.85	166	24.92	147.85	152	53.17	146.33	139.5	85.41	135.72
−20	0	19°	34.43	165	26.8	158.90	151	55.43	152.56	138	90.04	143.07
−10	0	29°	55.43	163	30.57	181.30	148	62.49	171.98	133	107.24	170.41
0	0	39°	80.98	161	34.43	204.20	143	75.36	207.40	128	127.99	203.38
10	0	49°	115.04	158	40.46	239.60	138	90.04	247.81	122	160.03	254.29
20	0	59°	166.43	153.61	49.62	294.25	131.6	112.63	309.99	114	224.60	356.89
23.44	0	62.44°	191.61	151.56	54.16	321.18	128.7	124.95	343.54	111.5	241.42	383.62

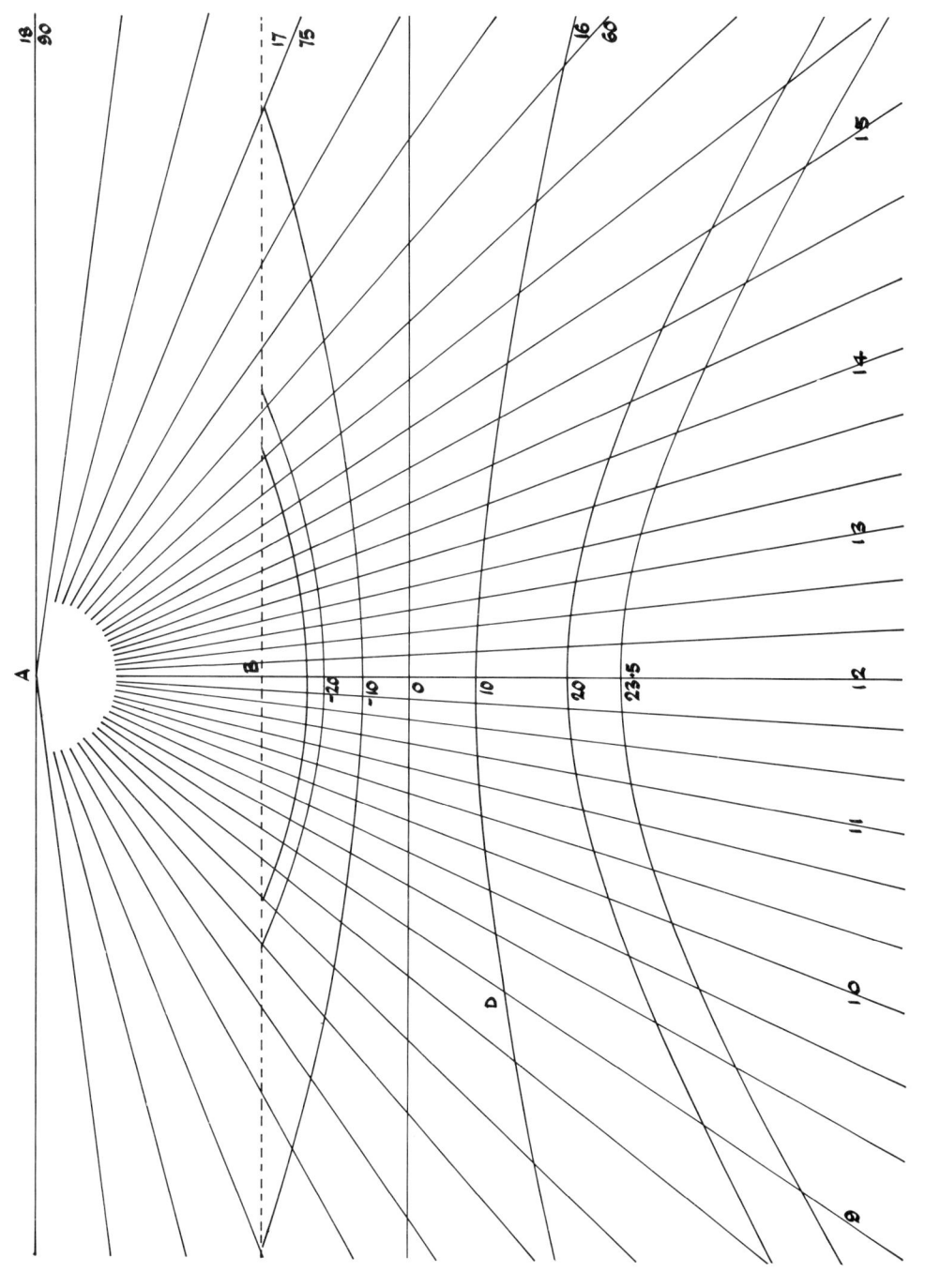

Fig. 7.23. Dial of a vertical south facing sundial showing declination tracks made by the shadow of the style point.

Positional Astronomy and Astro-Navigation Made Easy

In Table 7.8 the azimuths were calculated from the relation [3.7(8)],

$$\cot(Az) = \frac{\cos\phi \tan\delta - \sin\phi \cos(HA)}{\sin(HA)}.$$

It is useful to check values for azimuths from a set of sight reduction tables. Zodiacal lines based on the figures in the table are given in Fig. 7.23.

Fig. 7.24. A sundial in the Close of Salisbury Cathedral (S. facing declining) showing declination lines.

7.21 The Declination Lines for a Horizontal Dial

In Fig. 7.25, h is the height of the nodus above the horizontal surface of the dial. C is the shadow of A the nodus, and C follows the declination path of the shadow. The coordinates of C are,

$$x = BC \sin \text{azimuth}$$

$$y = BC \cos \text{azimuth}$$

Projects with Sundials and the Calculator

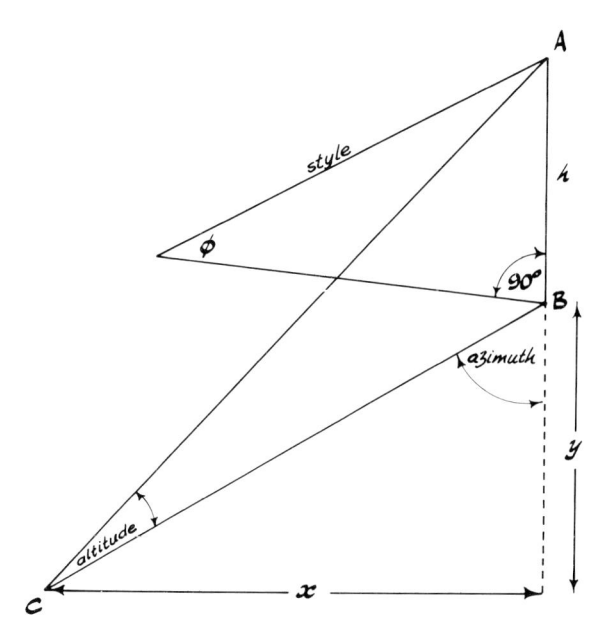

Fig. 7.25. Dial of a horizontal sundial showing declination tracks made by the shadow of the style point. B is the perpendicular projection of the style point. Height, nodus to $B = 5.06$ cm.

but,
$$\frac{h}{BC} = \tan(\text{alt}).$$

We find the (alt) from,
$$\sin(\text{alt}) = \sin \phi \sin \delta + \cos \phi \cos \delta \cos(\text{HA})$$

The azimuth is then given by,
$$\sin(\text{Az}) = \frac{\sin(\text{HA}) \cos \delta}{\cos(\text{alt})}.$$

From this x and y follow.

By way of example one point will be considered, namely $\delta = -10$, HA 30, latitude 51°.

$$\sin(\text{alt}) = \sin 51° \sin(-10°) + \cos 51° \cos(-10°) \cos 30°$$

and altitude = $23°.7$.

$$\sin(\text{Az}) = \frac{\sin 30° \cos -10°}{\cos 23°.7}$$

whence azimuth = $32°.53$.   (S $32°.53$ W)

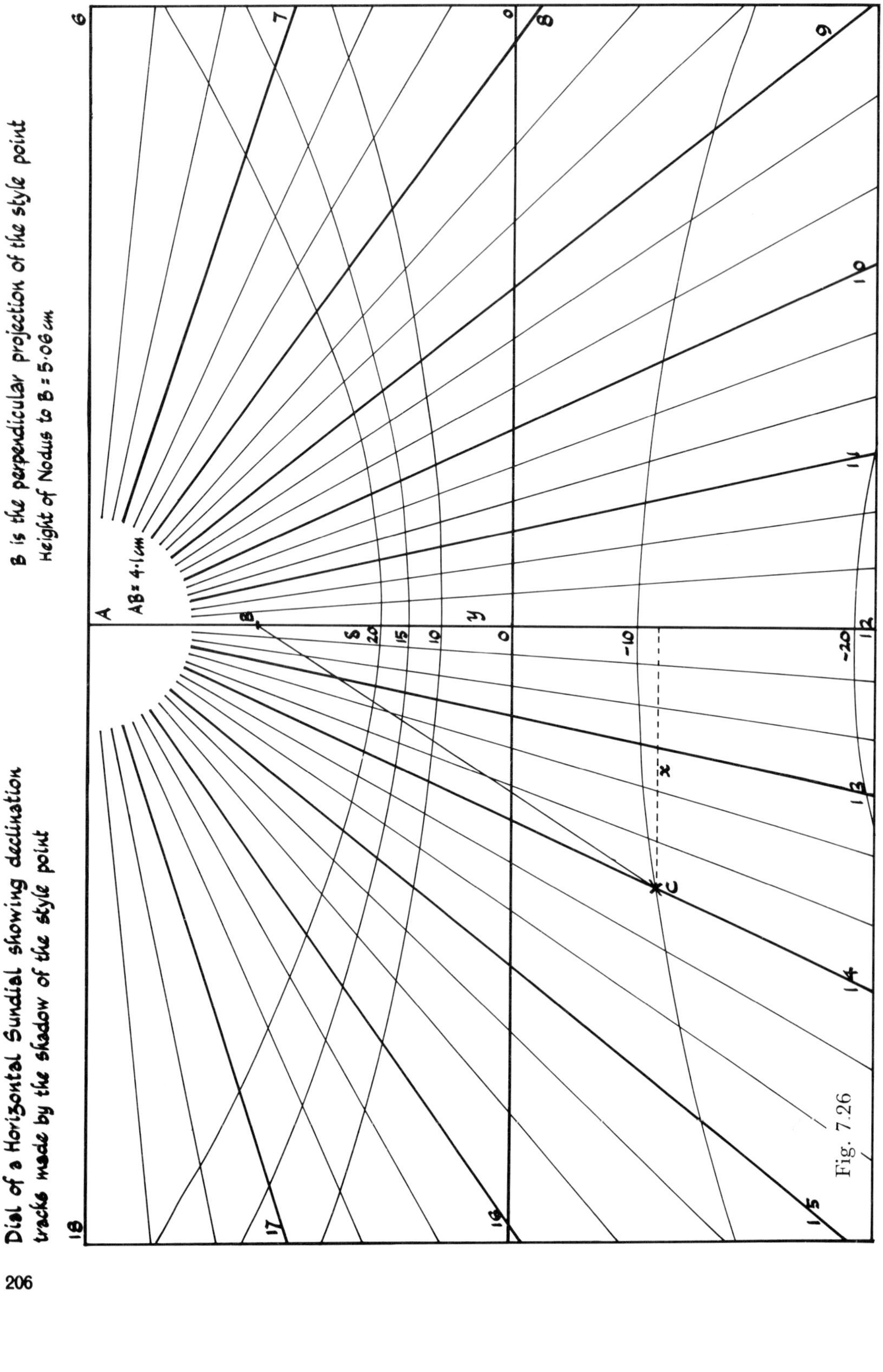
Dial of a Horizontal Sundial showing declination tracks made by the shadow of the style point
B is the perpendicular projection of the style point
Height of Nodus to B = 5·06 cm
AB = 4·1cm
A
B
C
x
y
S
20
15
10
0
-10
-20
6
7
8
9
10
11
12
13
14
15
16
17
18

Fig. 7.26

Projects with Sundials and the Calculator

Therefore
$$x = BC \sin 32°.53$$
$$y = BC \cos 32°.53.$$

In the graph for the horizontal dial $BC = \dfrac{h}{\tan(\text{alt})} = \dfrac{50°.6}{\tan 23°.7} = 115.$

$$x = 61.80 \qquad y = 96.95$$

which determines the point.

The locus of C can be calculated in this way for declinations 20°, 15°, 0°, −10°, −20°, as shown in Fig. 7.26.

8

The Equation of Time

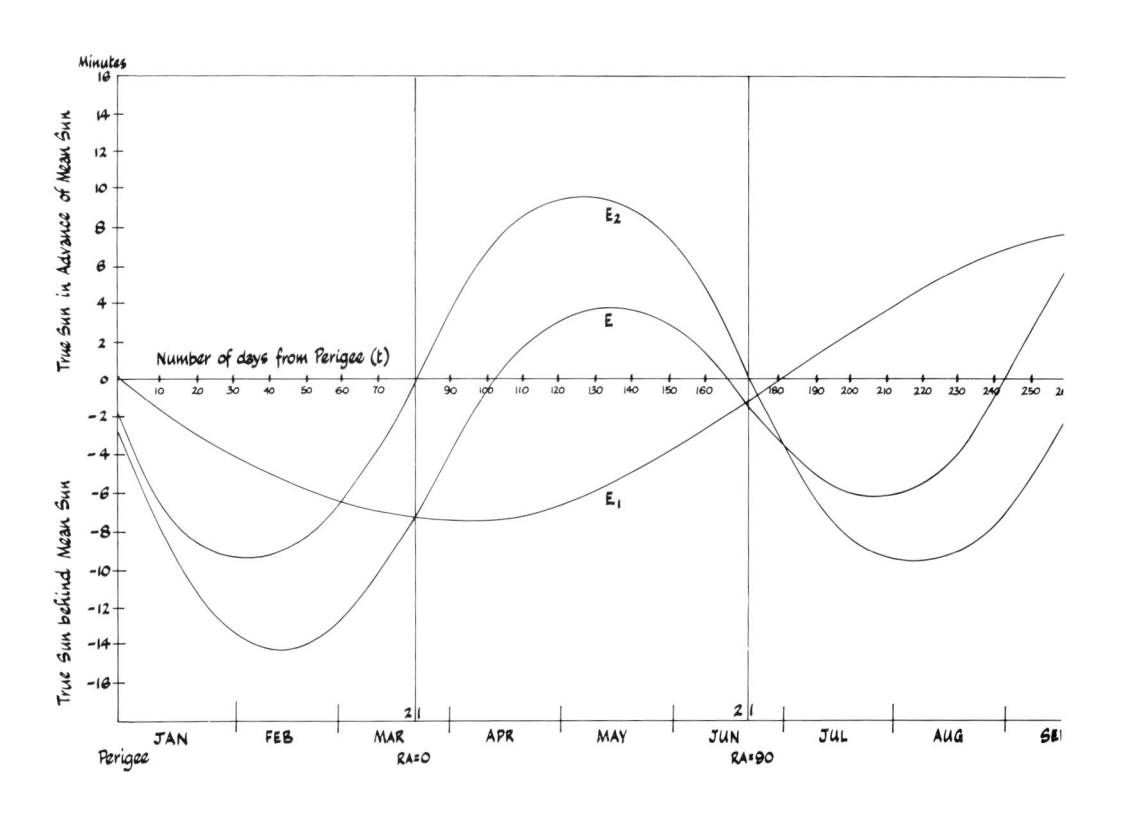

Positional Astronomy and Astro-Navigation Made Easy

8.1 The Equation of Time

Until the advent of atomic clocks with an accuracy of 1 part in 10^{13}, our time measurements were based on positional astronomy. Since we used to set our clocks by referring to the Earth's period of rotation we defeated attempts to time the period of the Earth's rotation. It is worth recalling the story of the factory manager before the days of time signals who used to set his watch every day when he passed a jeweller's shop, by the impressive clock in the window. The manager's watch was of course the factory standard. One day he happened to meet the jeweller and expressed his gratitude to the jeweller for his useful time keeping service. The jeweller became a little embarrassed and confessed that he always regulated his window clock by the factory hooter!

The subject of 'time' in all its astronomical aspects is a matter for advanced works on positional astronomy and for reference to the *Explanatory Supplement to the Astronomical Ephemeris* (HMSO).

We now know that the Earth's period of rotation is subject to a number of small irregularities. We have seen (Section 5.10) that sidereal time is based on the interval between two successive transits by a star of the observer's meridian. This way of reckoning time is impracticable for our day-to-day use, as the sun regulates our daily lives, our clocks and our calendars.

The interval between two successive transits of the Sun across the observer's meridian is a solar day, but this interval is not the same throughout the year. The '*mean* solar day' is therefore used, and is simply the mean of all solar days of the year. It is also called 24 hours of 'mean solar time'. For the practical purposes of this book the mean solar time and universal time are identical.

8.2

In this chapter we deal with the easily observed irregular habits of the 'real sun' as observed from the Earth. These irregularities are brought about by the fact that the Sun, Earth and Moon form a dynamic system. In this system the Earth travels round the Sun in an elliptical orbit with an eccentricity of about 1/60 in the plane of the ecliptic.

Most amateur astronomers, navigators and sundialists are familiar with the fact that the Sun transits the meridian on some days several minutes ahead of 12 noon local mean time and on other days of the year it transits several minutes after 12 noon. That is, the *mean sun* and the *true sun* are out of step and the amount by which the true sun is ahead of the mean sun, in

The Equation of Time

minutes and seconds of *mean time*, is known as the *equation of time*, denoted by E (see Section 2.8).

For simple practical observations E is the amount of time that must be added to, or subtracted from, UT to get the true sun time. E is given for each day in the *Nautical Almanac* and in *Whitaker's Almanack*, and can be regarded as the excess of the Right Ascension of the mean sun over the true sun, or the excess of the LHA of the true sun over the mean sun.

8.3 There are two factors mainly responsible for the equation of time.

(1) The Earth moves round the Sun in an elliptical orbit. This factor is ascribed to the eccentricity of the Earth's orbit.

(2) The plane of the Earth's orbit is inclined to the plane of the Earth's equator. This factor is ascribed to the obliquity of the ecliptic.

When mathematicians and astronomers set to work to find a relation that would express this small corrective amount of time in terms of the factors that might affect it, such as the time of the year, the Sun's longitude and the obliquity of the ecliptic, one useful approach to this problem was to consider separately the two main factors that in combination appeared to produce the Sun's time keeping irregularities as observed from the Earth.

By this means an expression can be established having several constants involving the eccentricity of the Earth's orbit and the obliquity of the ecliptic, and with the Sun's longitude as the only variable.

It is this expression which can provide a value for E in minutes and seconds.

In the foregoing sections some simple methods are suggested for the satisfaction of actually calculating what the equation of time is. This provides scope for using the calculator on some projects concerning time. Accurate daily values for the equation of time are readily available for astronomers and navigators from the *Nautical Almanac* or the *Astronomical Ephemerides*.

The component of the irregularity caused by the eccentricity of the Earth's orbit is denoted by E_1 and the component produced by the obliquity of the ecliptic is denoted by E_2, so that,

$$E = \text{the excess of true sun time of GMT or UT}$$
$$= E_1 + E_2.$$

Values of E_1 and E_2 can be either positive or negative.

8.4 The Eccentricity of the Earth's Orbit, e — How it can be Measured

It is a matter of interest and requires only a simple calculation to verify the eccentricity of the Earth's orbit. This can be done by first finding the ratio of the distances (focus to perigee) : (focus to apogee). From Fig. 8.1 in which E is the position of the Earth, this ratio is $EA : EA'$. The distances EA and EA' are inversely proportional to the Sun's apparent diameter at these distances. The Sun when at perigee as observed from E will appear to have a slightly larger diameter than when the moon is at A apogee. This can be measured with a good telescope, but for our present purpose we can take the minimum and maximum values given in *Whitaker's Almanack* or in the *Nautical Almanac*, and these are, semi-diameter at $A = 16'18''$, semi-diameter at $A' = 15'45''$.

The eccentricity, e, of the ellipse can be defined as

$$\frac{CE}{CA} = e \qquad CE = eCA.$$

The semi-diameters are in the inverse ratio of the distances of the Sun from the Earth, so,

$$\frac{EA'}{EA} = \frac{16'18''}{15'45''} \qquad \frac{CA+CE}{CA-CE} \qquad CE = eCA$$

Therefore

$$\frac{EA'}{EA} = \frac{CA+eCA}{CA-eCA} = \frac{CA(1+e)}{CA(1-e)}$$

$$= \frac{1+e}{1-e} = \frac{16.3}{15.75}$$

$$e = \frac{16.3 - 15.75}{16.3 + 15.75}$$

$$e = 0.017\ 16 = 1/60 \text{ (approx.)}.$$

The calculator makes light work of this calculation.

8.5

In Fig. 8.1, $ABA'B'$ represents the Earth's orbit round the Sun. It is an ellipse of small eccentricity.

This is also the apparent orbit of the true sun round the Earth. In this account the 'sun' refers to the apparent sun as observed from the Earth.

E and H are the foci of the ellipse, and C is its centre. $CA = CA'$.

The Equation of Time

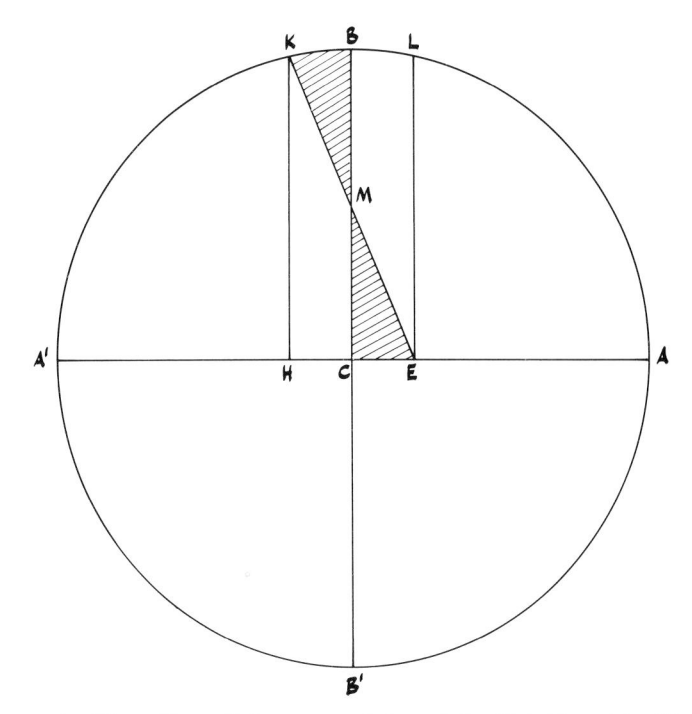

Fig. 8.1. The effect of the eccentricity of the Earth's orbit. The eccentricity e of the ellipse can be defined as $CE/CA = e$.

The mean sun that gives us our time apparently travels uniformly round a path in the plane of the Earth's equator and makes equal angles in equal times. At L the Sun appears to have gone through $90°$.

The apparent true sun, however, has travelled faster because by Kepler's 2nd Law it describes equal *areas* in equal times.

So while the mean sun goes from A to L, the true sun goes from A to K, whence,

$$KMB = CEM.$$

It describes the area $AKEA$ in a $\frac{1}{4}$ of the periodic time and $AKEA$ is $\frac{1}{4}$ the area enclosed by the orbit.

$ABCA$ is also $\frac{1}{4}$ the area of the orbit since KMB is approximately equal to CEM.

The true sun has moved through an angle KEL more than the mean sun.

$$\tan KEL = 2/60 \quad \text{and} \quad KEL = 1°.91.$$

This is the difference in RA between true sun and the mean sun. $1°$ in (RA) $= 4$ min, so in the situation shown in Fig. 8.1 the RA true sun is 7.64^{m} ahead of the mean sun, or $7^{m}\,38^{s}.4$. See 8.2.

213

It is apparent from Fig. 8.1 that as the Earth moves fastest round the Sun when it is nearest the Sun (at perigee A), the Sun will appear to gain on the apparent mean sun—which by definition moves at a uniform *angular* velocity in the plane of the equator. The true sun and the mean sun, however, appear level again at A, and this means that the true sun must move slowly in its orbit after reaching the half-way mark, i.e., when its longitude is 90°. The difference between 'true sun time' and the 'mean dynamical sun' by Fig. 8.1 is a maximum at 90° longitude, for E_1.

It is reasonable to assume that the maximum difference in position, and therefore in time, will occur when $AEL = 90°$. This suggests that the value of this difference in position varies sinusoidally from 0° to 180° from *2nd January to 2nd July (perigee to apogee)*.

The angle to be used in the sine relation for any date is ωt where ω is the uniform angular velocity of the Earth in its orbit, and t is the number of days that have elapsed from 2nd January to the date for which the advance or retardation in time is required. We can make approximate calculations using this simplified principle.

The advance in HA is therefore $-7.637 \sin \omega t$. (from above). ω for practical purposes is 360° in 365.25 days = 360/365.25.

The expression shows that on 2nd January $E_1 = 0$

When $\omega t = 90$	about 1st April	$E_1 = -7^m.637$ or $-7^m 38^s.4$
$\omega t = 180$	about 2nd July	$E_1 = 0$
	and on 1st October	$E_1 = -7^m.637$ or $-7^m 38^s.4$

The effect of the eccentricity of the Earth's orbit on the Sun's time keeping (E_1) can best be shown on a graph for $E_1 = -7^m.637 \sin \omega t$ (Fig. 8.3).

8.6 The Component of E Caused by the Obliquity of the Ecliptic

Quite separate from the irregularity of the Sun's time keeping due to the ellipticity of the Earth's orbit round the Sun, is the irregularity caused by the fact that the Sun appears to move during the course of a year in a path on the ecliptic which is inclined to the equator at an angle of 23°.44, which has the symbol ε. The equator is really the basis of our time measurements. Our daily time system is based on the Earth's revolution round the polar axis, and the equator is the plane at right angles to this axis.

The Equation of Time

It can be seen from the Fig. 8.2 that during March and September, i.e., at ♈ and ♎, the Sun, during one day of progress along the ecliptic, will travel 1°, but along the equator this will mark out only $1° \cos \varepsilon$ (where $\varepsilon = 23°.44$), i.e., $1° \cos 23°.44 = 0.917\,477$.

The Sun during this day will thus lose $1.0 - 0.917\,477 = 0°.082\,58$ of RA but will gain $0°.082\,58$ of HA because RA and HA are measured in opposite directions.

In Fig. 8.2, S is the true sun, S_1 the dynamical mean sun which appears to move in its annual motion in the plane of the ecliptic, and S_2 is a point on the equator so that $♈S_2 = ♈S_1$.

S_2 is the *astronomical mean sun* with apparent motion in the equatorial plane. E_2 is the part of the equation of time due to the obliquity of the ecliptic. $E_2 =$ the difference in hour angle between S_1 and $S_2 =$ angle $S_1 P S$.

This can be regarded as the advance of the 'ecliptic mean sun' over the 'equatorial mean sun', or the advance of the 'dynamical mean sun' over the 'astronomical mean sun'. The dynamical mean sun in the ecliptic, and the astronomical mean sun in the plane of the equator are *in step* when $♈S_1 = ♈M$.

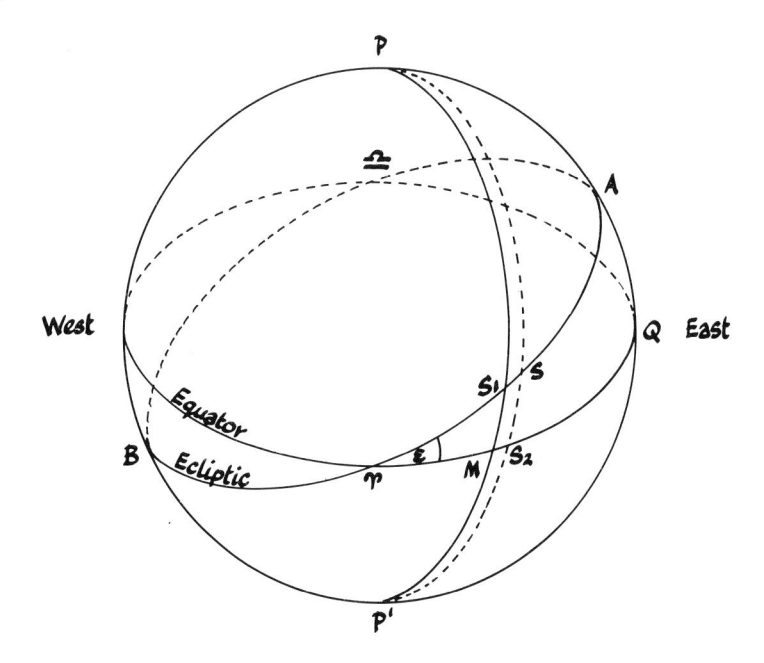

Fig. 8.2

8.7

We have so far merely made a rough qualitative estimate of the effect of the obliquity of the ecliptic (E_2) on the equation of time.

We treated a one day apparent movement of the Sun from ♈ as making a small plane triangle with the equator; while this is not greatly in error for a 1° movement from ♈, it will be badly in error for a time of one month. We must regard ♈ $S_1 M$ as a *spherical* triangle (Fig. 8.2).

If we apply the four part relation [3.7(8)] then,

$$\cos ♈M \cos \varepsilon = \sin ♈M \cot ♈S - \sin \varepsilon \cot S_1 M ♈ \quad \text{(where } \varepsilon = 23°.44\text{)}$$

but $S_1 M ♈ = 90°$ and $\cot S_1 M ♈ = 0$,

therefore $\quad \tan ♈S_1 \cos \varepsilon = \tan ♈M$ [8.7(1)]

$\cos \varepsilon$ is a constant and $= 0.91747$.

The dynamical mean sun (in the ecliptic) and the astronomical mean sun (in the equator) are in step when ♈S_1 = ♈M, i.e., when they both equal 0, 90, 180 and 270. There would appear to be a maximum difference between the two when ♈$S_1 \approx 45°$ = longitude of the mean dynamical sun, and this can be checked using a calculator. From relation [8.7(1)],

$$\tan ♈M = 0.91747 \tan 45$$

$$♈M = 42°.536$$

then $\quad E_2 = ♈S_1 - ♈M = 45 - 42.536 = 2°.464 \ (1° = 4 \text{ minutes})$

$\quad\quad\quad\quad\quad = 9.856 \text{ minutes}.$

We estimated the maximum value might occur when the longitude of the mean sun was 45°. This is very near the longitude for maximum E_2, but it can be shown by trying longitude 46° and then 47° that E_2 (max) occurs at longitude 46°. There are doubtless more rigorous ways of finding the Sun's longitude for E_2 (max) but this simple calculator check is adequate for our purpose.

As E_2 varies approximately sinusoidally with 2(RA), we may take the value of E_2 to be given by

$$E_2 = E_2 \text{ max} \sin 2(\text{RA}) \quad [8.7(2)]$$

$$E_2 = 9.863 \sin 2(\text{RA}) \quad [8.7(3)]$$

The factor of 2 in 2(RA) is required as $E_2 = 0$ when (RA) = 0°, 90° 180° or 270° and is a maximum when $\sin 2(\text{RA}) = 1$, i.e., when (RA) = 45° or 135°.

E_2 changes sign four times in the course of one year and E_1 changes sign only twice in a year, as is shown in Fig. 8.3.

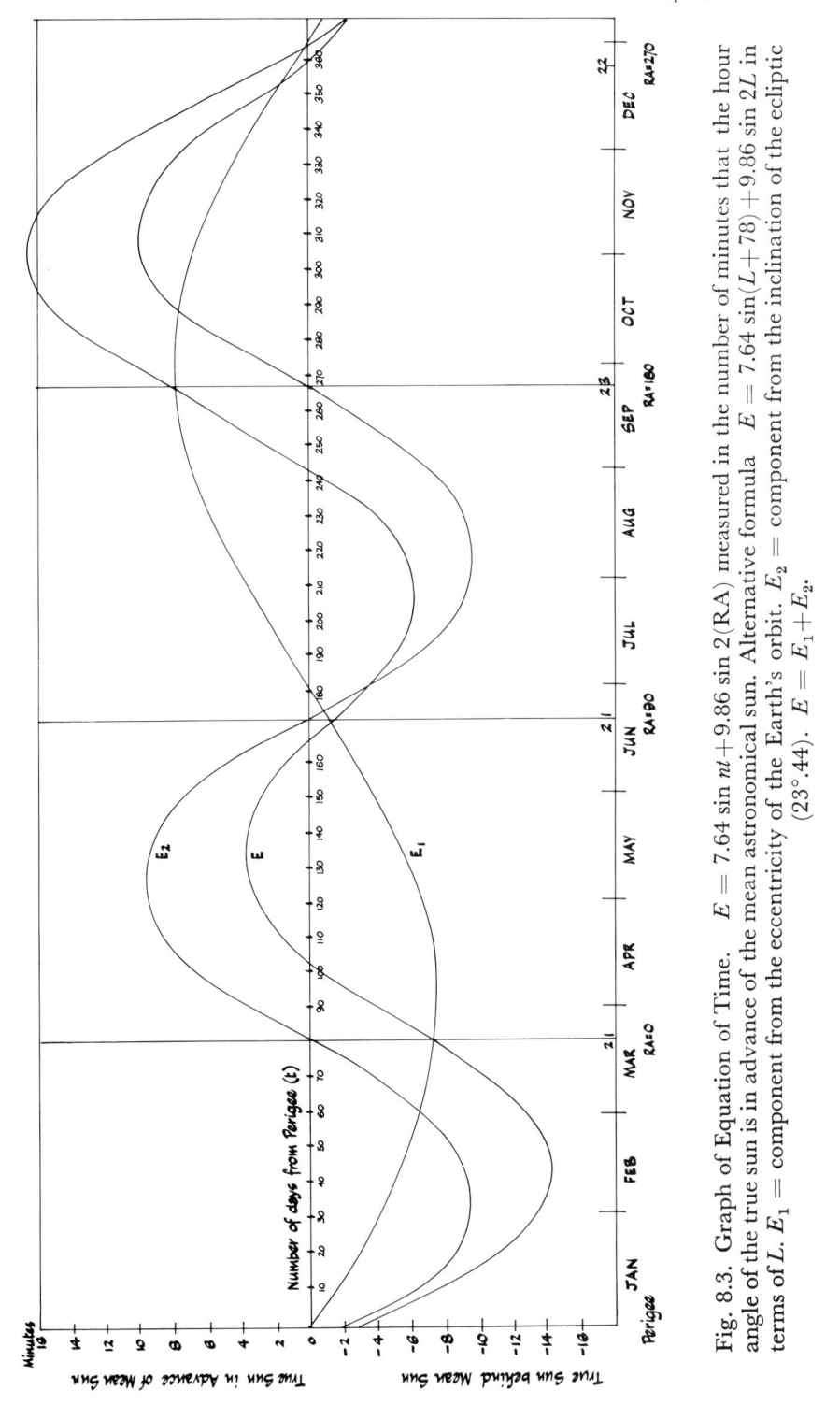

The Equation of Time

Fig. 8.3. Graph of Equation of Time. $E = 7.64 \sin nt + 9.86 \sin 2(\text{RA})$ measured in the number of minutes that the hour angle of the true sun is in advance of the mean astronomical sun. Alternative formula $E = 7.64 \sin(L+78) + 9.86 \sin 2L$ in terms of L. $E_1 =$ component from the eccentricity of the Earth's orbit. $E_2 =$ component from the inclination of the ecliptic (23°.44). $E = E_1 + E_2$.

Positional Astronomy and Astro-Navigation Made Easy

8.8 We now have a workable approximate formula for finding the equation of time from the date and the Sun's RA on that date, using relations [8.7(2)] and [8.7(3)]

$$E = E_1 + E_2$$
$$= -7.64 \sin \omega t + 9.863 \sin 2(\text{RA}). \qquad [8.8(1)]$$

It is a rewarding and instructive exercise for the calculator to plot three graphs from the results obtained from the separate components of Fig. 8.3.

$$E_1 = -7.64 \sin \omega t$$
$$= -7.64 \sin \left(\frac{360}{365.25} \times \text{number of days since 1st January} \right)$$

The graph E_1 is shown in the Fig. 8.3.

Plotting E_2 from $E_2 = 9.864 \sin (2 \times (\text{RA}))$ is straightforward and the graph is also shown in Fig. 8.3.

E is simply $E_1 + E_2$ and shows the familiar graph for the equation of time.

The values for E can be checked from the *Nautical Almanac* or from *Whitaker's Almanack*.

8.9 The equation of time is sometimes expressed in a slightly more simplified form in terms of the Sun's longitude alone instead of in terms of ω and the RA, but a little calculating will show that there is a difference of only a few seconds between the two formulae.

We use in place of the RA of the Sun the ecliptic coordinate or longitude of the Sun, to be found in tables, e.g., in *Norton's Star Atlas*. The longitude and RA of the Sun are approximately the same value for the purposes of the equation. It will be noticed that the component E_1 has its starting point or zero on 2nd January (apogee), but E_2 has its zero on 21st March. There is thus a difference of 'phase' of approximately 79 days which represents 78° of longitude.

So we can use the Sun's longitude L as follows:

$$E = -7.64 \sin(L+78) + 9.863 \sin 2L$$

which involves only the Sun's longitude.

The Equation of Time

8.10 Examples

(1) Find E on 12th February by longitude of Sun.

$$\text{Longitude of Sun} = 322$$
$$\phantom{\text{Longitude of Sun} = }\underline{78}$$
$$\phantom{\text{Longitude of Sun} = }400$$

$$E = -7.64 \sin 400 + 9.86 \sin 644$$
$$= -14.4780 = -14^m\ 28^s.6$$

which is an approximate value.

(The calculator handles sin 400 and sin 644 with great ease and with the correct sign.)

(2) To find E for 21st August, longitude 147, applying formula
$$E = -7.64 \sin(147+78) + 9.864 \sin(2 \times 147)$$
$$= -3.61.$$

(3) To find E for 31st July, longitude 127.54.
$$E = -7.64 \sin 205.53 + 9.864 \sin 255.08$$
$$= -6.237\ 53 = 6^m\ 14^s.$$

(4) To find E for 31st October by the Day and RA method
$$E = -7.64 \sin 297.65 + 9.864 \sin 430.675$$
$$= 16.075 = 16^m\ 4^s.5.$$

By Longitude $= 217$
$$E = -7.64 \sin 295 + 9.864 \sin 434$$
$$= -16.406 = 16^m\ 24^s.36.$$

The method by longitude is the better in this case.

(5) Calculate the 'equation of time' on 1st November 1976 when $n = 302$.
$$E_1 = -7.64 \sin nt$$
$$= -7.64 \times \sin \frac{360}{365.25} \times 302$$
$$= -7.64 \sin 297.66$$
$$= -6.767 \text{ min.}$$
$$E_2 = 9.864 \sin 432.624$$
$$= 9.4147 \text{ min.}$$
$$E = E_1 + E_2 = -16.1817 \text{ min} = -16^m\ 11^s.$$

Having found approximate expressions for giving values of E_1 and E_2 in terms of the Sun's longitude, the calculator will enable you to draw up a table, and to plot the graphs in Fig. 8.3.

Showing longitudes at regular intervals throughout the year and the corresponding values for E_1 and E_2 and also E.

There are various ways of representing the time along the abscissa, either by the Sun's longitude which varies with the time of year, or by the calendar dates.

As an aid in drawing each curve it is useful to mark the maxima for E_1, namely -7.64 minutes at about 2nd April and $+7.64$ at end of September. Maxima for E_2, 9.86 minutes, as we have found, occur at 1st February $(-)$, 1st May $(+)$, 4th August $(-)$ and 4th November $(+)$.

8.11 Noon Marks

Although a reference to 'noon marks' would perhaps have been more appropriate under the section on sundials, it is introduced here under the equation of time as an 'analemmatic noon mark' is a means of indicating on a sundial the variations in the equation of time by showing the mean time of the true sun's transit of the meridian throughout the year.

In Fig. 8.4, *CDEF* is a south facing vertical dial with a style which has a nodus which projects perpendicularly a distance of 100 mm from the vertical plane of the dial. If the shadow, *H*, of such a nodus be marked at local mean noon each day of the year the marks will form the pattern shown, which is approximately symmetrical about the perpendicular, *AK*. This pattern is really an alternative way of plotting the values of the equation of time throughout the year, and conveys the same information as the graph for E in Fig. 8.3. A curve formed by plotting the equation of time along the x axis and the Sun's declination throughout the year has the shape of a figure of eight and is slightly asymmetrical about the prime meridian line (Fig. 8.5).

The way in which the coordinates for an analemmatic noon mark can be calculated is best explained by considering the position of the noon shadow of a nodus on a particular day.

For example, consider the shadow of the nodus *B* formed at mean time noon on 1st November. This is marked as 1st Nov. in Fig. 8.4.

From the tables of E or by the calculation in Section 8.9 we find that the Sun's HA at mean noon on this day is $4°.0915$ ($= 16^m\ 5^s.5$).

The declination of the Sun is $-14°.4$ (from *Whitaker's Almanack* or *Nautical Almanac*).

The Equation of Time

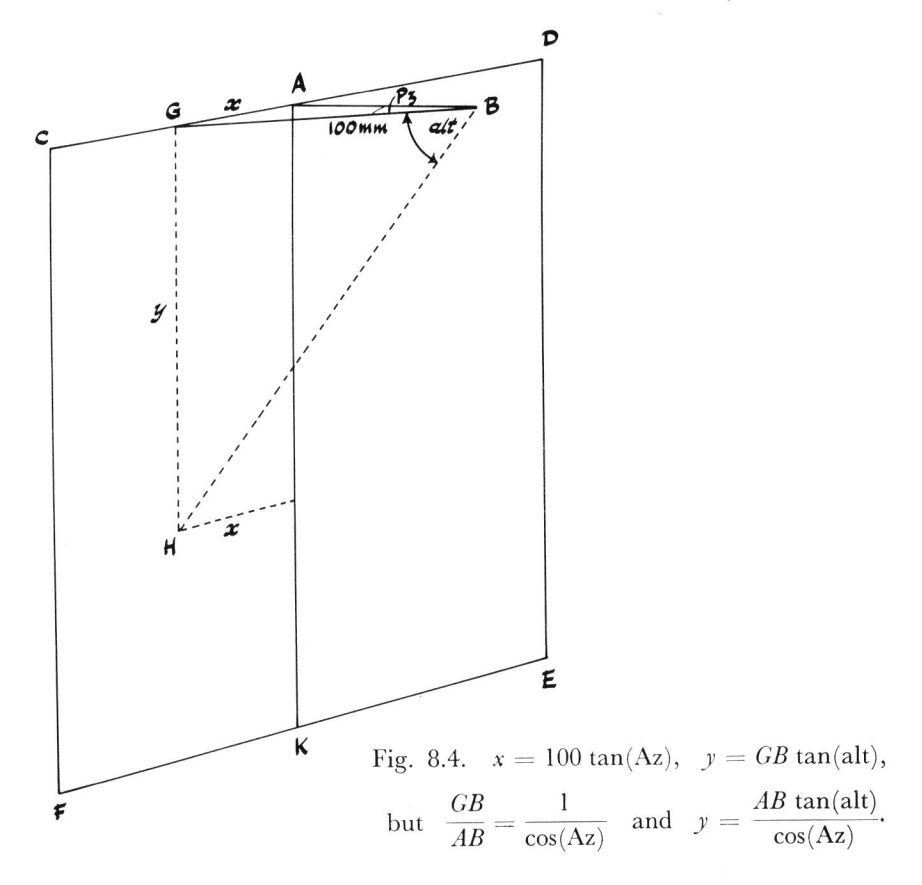

Fig. 8.4. $x = 100\tan(\text{Az})$, $y = GB\tan(\text{alt})$, but $\dfrac{GB}{AB} = \dfrac{1}{\cos(\text{Az})}$ and $y = \dfrac{AB\tan(\text{alt})}{\cos(\text{Az})}$.

The approximate altitude for lat $51° = (90-51+\delta) = 24°.6$.

The azimuth of the Sun, from

$$\sin(\text{Az}) = \frac{\sin(\text{HA})\cos}{\cos(\text{alt})} = +4°.359$$

From the geometry of Figure 8.5, we have,

$$x = 100\tan(\text{Az})$$

$$y = \frac{100\tan(\text{alt})}{\cos(\text{Az})}$$

Using the azimuth and altitude above,

$$x = 100\tan 4°.359 = 7.6227 \text{ mm}$$

$$y = \frac{100\tan 24°.6}{\cos 4°.359} = 45.9 \text{ mm}$$

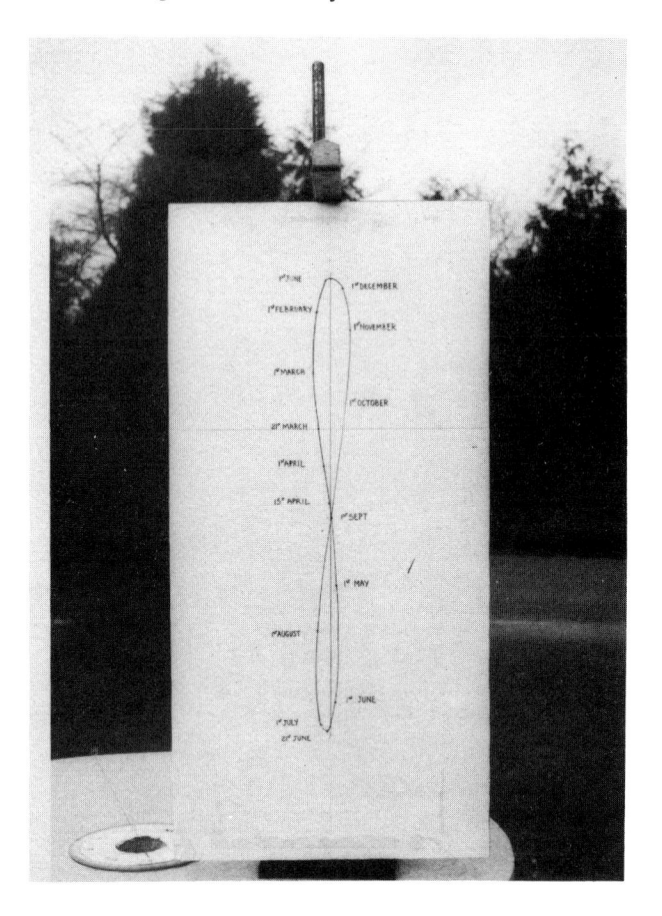

Positional Astronomy and Astro-Navigation Made Easy

Fig. 8.5

We can therefore plot the point on a sheet of graph paper, corresponding to the noon mark of B at local mean noon on 1st November.

The coordinates for the noon mark for a vertical dial are given in Table 8.1. From this table the analemmatic noon marks of Fig. 8.6 were drawn.

The scale can easily be changed for making a noon mark suitable for use in a garden or on the side of a house by increasing the scale, e.g. making the style 300 mm instead of 100 mm. See Fig. 8.5.

8.12 The Polar Dial—Noon Mark

In Section 7.16 a brief account was given of a polar sundial and the co-ordinates of the spot where the nodule of the style would fall at various times of day and day of the year were caculated from the relations, directly applicable to the polar dial.

The Equation of Time

$$x = \text{length of style} \tan H$$

$$y = \text{length of style} \frac{\tan \delta}{\cos H}.$$

Table 8.1. Noon marks for a vertical dial.
Length of horizontal style is 100 mm.

	Sun's HA at mean noon (HA) = E	δ	Altitude $39+\delta$	Azimuth Z from $\sin(Az) = \frac{\sin(HA)\cos\delta}{\cos(alt)}$	$x = 100\tan Z$	$y = 100\frac{\tan(alt)}{\cos Z}$
	°	°	°	°	in mm	in mm
1st Jan	−0.7582	−23.083	15.917	0.725	−1.26	28.52
1st Feb	−3.3665	−17.4	21.6	3.455	−6.037	3.966
1st Mar	−3.1125	−7.6166	31.383	3.6143	−6.3168	61.121
21st Mar	−1.825	0	39	2.348	−4.100	81.046
1st Apr	−0.9958	+4.50	43.5	1.3686	−2.309	94.923
15th Apr	0	+9.733	48.733	0	0	113.96
1st May	+0.725	+15.05	54.05	1.1926	+2.0818	137.921
1st Jun	+0.57075	+22.033	61.033	1.0925	+1.907	180.68
21st Jun	−0.375	+23.44	62.44	−0.7436	−1.298	191.624
1st Jul	−0.92915	+23.1166	62.116	−1.8274	−3.1905	189.09
1st Aug	−1.5708	+18.0333	57.033	−2.7456	−4.796	154.358
1st Sept	−0.0165	+ 8.3166	47.317	−0.024 082	−0.004 203	108.433
21st Sept	+1.875	0	39	2.429	+4.242	81.05
1st Oct	+2.558	−3.15	35.85	+3.1516	+5.506	72.364
1st Nov	+4.0915	−14.4	24.6	+4.359	+7.6226	45.916
1st Dec	+2.7583	−21.7833	17.2167	2.6814	+4.6833	31.021

The polar dial (see Section 7.6) has a special simplicity about it, and a table of coordinates for noon mark can easily be constructed using a calculator.

In order to plot the curve for the noon marks (Fig. 8.6) we can take the first 3 columns of the tables for the vertical 'noon mark', Table 8.1, as these are common, and complete it for the polar dial shown in Table 8.2.

The table for the coordinates measured from the base of the style (height 100 mm) is as in Table 8.2 and is similar to that for a vertical dial.

Table 8.2 contains only a few points, but readers will find a smooth curve will result from a few more points.

A noon mark for a polar dial is shown, Fig. 8.7. The nodus in this case is 200 mm (Fig. 8.3). The meticulus and complicated geometrical construction for these curves is well described in books on sundials, for example in *Sundials* by Frank Cousins, but the calculator has brought the tedious and exacting tasks described well within the capabilities of 'A' level students, as the tables and graphs show.

Table 8.2. Noon marks for a polar dial. Height of nodus is 100 mm.

	Sun's HA at mean noon (HA) = E	δ	$x = 100\tan(\text{HA})$	$y = 100\,\dfrac{\tan\delta}{\cos(\text{HA})}$
	°	°	in mm	in mm
1st January	−0.7582	−23.083	−1.323	+42.622
1st February	−3.3665	−17.4	−5.824	+31.392
1st March	−3.1125	−7.6166	−5.438	+13.39
21st March	−1.825	0	−3.186	0
1st April	−0.9958	+4.50	−1.738	−7.87
1st May	+0.725	+15.05	+1.265	+26.89
1st June	+0.570 75	+22.033	+0.996	−40.47
21st June	−0.375	+23.44	−0.654	43.357
1st July	−0.929 15	+23.1166	−1.622	−42.69
1st August	−1.5708	+18.0333	−2.742	−32.56
1st September	−0.0165	+8.3166	−0.029	−14.62
21st September	+1.875	0	+3.275	0
1st October	+2.558	−3.15	+4.467	+5.51
1st November	+4.0915	−14.4	+7.153	+25.68
1st December	+2.7583	−21.7833	+4.818	+39.97
21st December	+0.50	−23.44	+0.872	+43.37

The Equation of Time

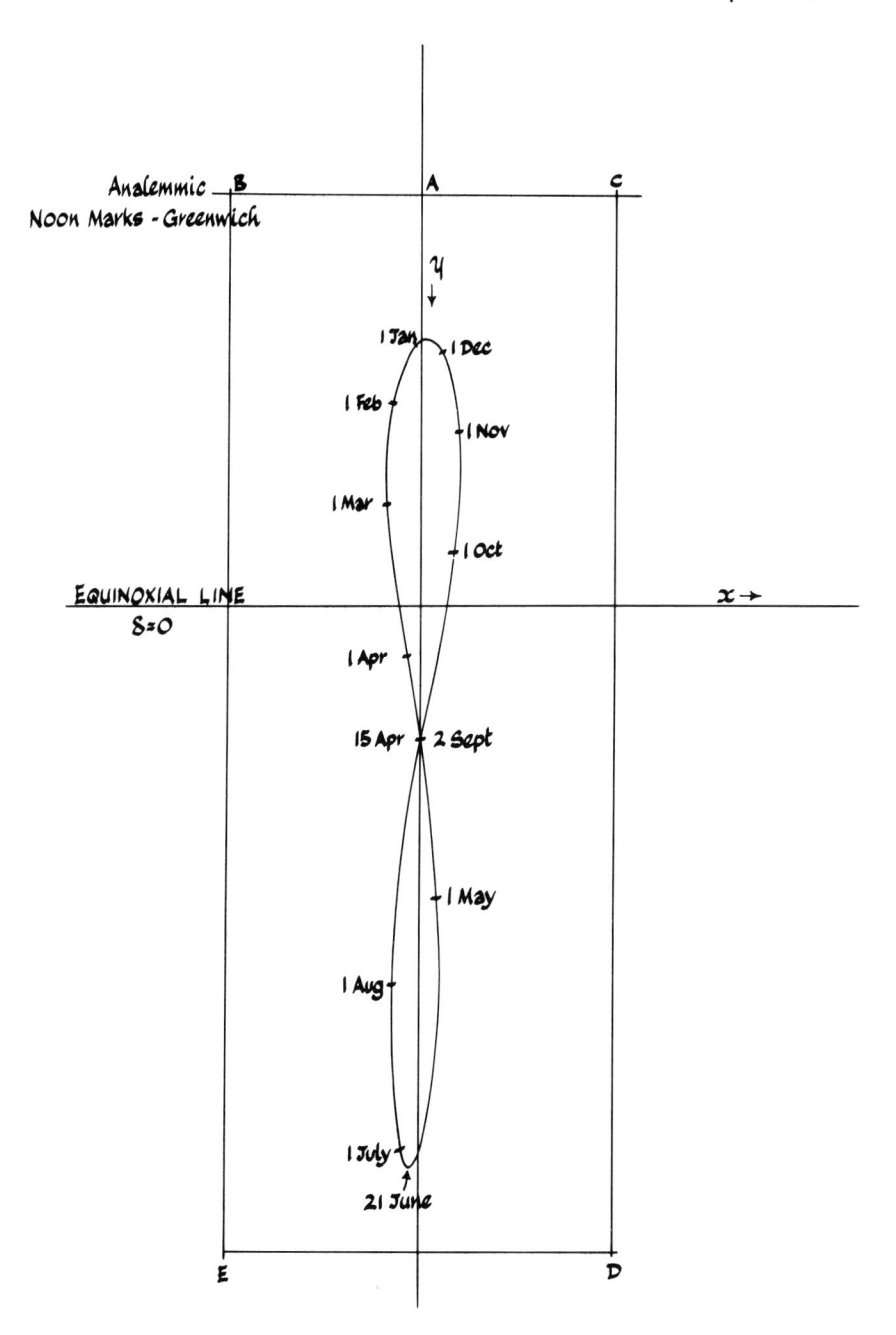

Fig. 8.6. Vertical dial, south facing, with nodus at A at perpendicular distance of 100 mm from plane of dial. Latitude 51°N. When the Sun casts a shadow of the nodus on the curve, this signifies it is local noon (mean time).

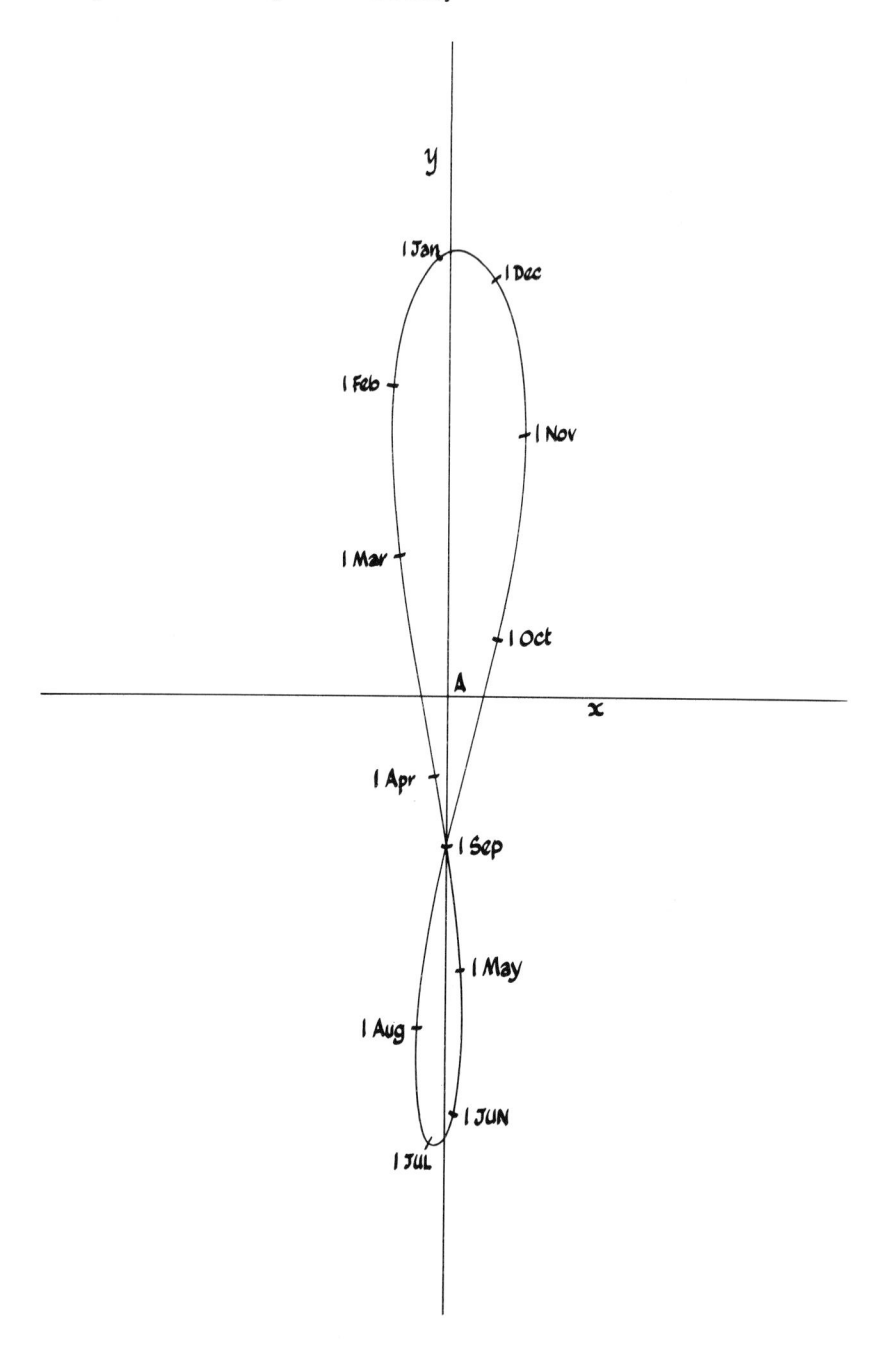

Fig. 8.7. Analemmic noon mark for a polar dial. Height of nodus 200 mm at *A*. This figure is derived from Table 8.2.

9
Precession of the Earth's Axis

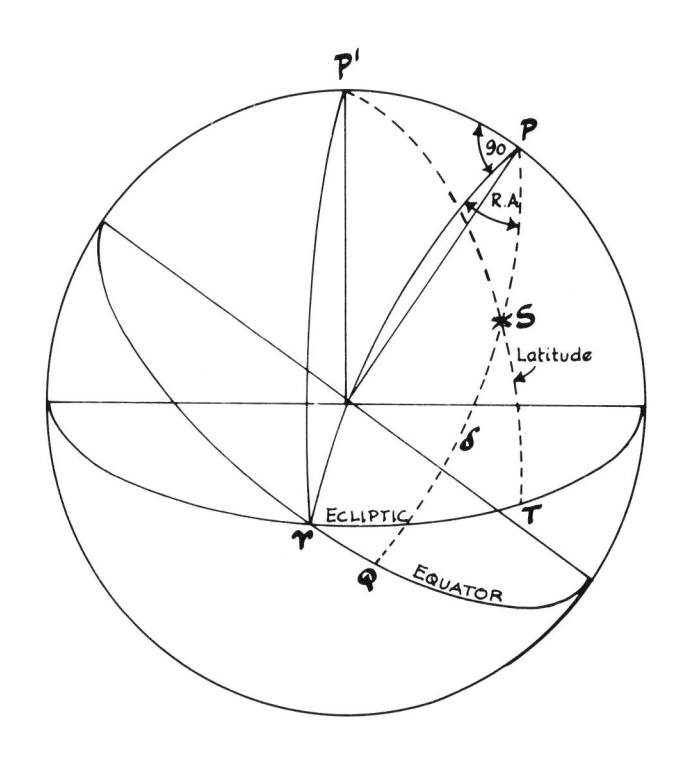

9.1

One of the fascinating facts of astronomy is that the Earth's axis precesses slowly round a fixed point in the celestial sphere rather as the axis of a spinning top will do when spinning at a small angle to the vertical.

This point in the celestial sphere has coordinates RA 18° and declination 66.5° North, and is called the 'North Pole of the ecliptic'.

The Earth's axis always points approximately $23\frac{1}{2}°$ away from the ecliptic pole and in the course of 26 000 years it describes a full circle round it (Fig. 9.1). During this circular tour of the heavens it points in succession to several bright stars. During this century it points to within 1° of Polaris in Ursa Minor but around the year 2000 BC, when the cultures of the Nile and Mesopotamia flourished, the Earth's axis pointed to alpha Draconis. Around the year AD 14 000 the Pole Star will be in the region of Vega. Despite this very slow precession, the changes in the point about which the heavens appeared to turn, was discovered by Hipparchus (c. 150 BC) who found that the celestial longitude of each star increased by about 50 seconds per year, but the celestial latitude of stars remained the same. This is because celestial latitude is measured from the plane of the ecliptic which itself is considered as being fixed.

As the Earth's pole makes its slow precession round the pole of the ecliptic, it causes the plane of the celestial equator to cross the ecliptic in a slightly different place each year. As we have seen in Section 2.3, the celestial equator crosses the plane of the ecliptic at two points known as the equinoxes and those points, as Hipparchus noticed, move round the ecliptic about 50 seconds each year so causing a corresponding increase in the longitude of all stars. This motion of the Earth's pole is an effect of the gravitational forces acting between the Sun and the Moon on the equatorial bulge of the Earth and is often referred to as the 'luni–solar precession'.

9.2

In this section we shall consider the relation between the two coordinate systems of stars:

(1) The equatorial system using Right Ascension and declination which refer to the Earth's axis pointing to the celestial pole, and the celestial equator.

(2) The system using the plane of the ecliptic as its reference by which stars are given coordinates, called celestial latitude and celestial longitude which are measured with reference to the plane of the ecliptic and the ecliptic pole (see Fig. 9.2). The star maps of Hipparchus used this system of latitudes and longitudes because the Moon and planets all appeared to follow approximately the Sun's track in the ecliptic. The Sun's celestial latitude is 0° as it remains in the plane of the ecliptic.

Precession of the Earth's Axis

The azimuths at rising and setting of stars and their altitudes at transit across the local meridian, depend on their declinations which change appreciably in the course of centuries owing to precession and swing through 47° (23.5°×2) every 12 900 years.

As a result of precession, as we have seen, the nodes or points of intersection of the Sun's path with the celestial equator shift in the course of a year by about 50 seconds of arc along the ecliptic circle in a direction opposite to the Sun's direction of annual retreat. The result of this is that every star (except those within $23\frac{1}{2}°$ of the poles) passes through every hour of Right Ascension from 0^h to 24^h once every 25 800 years.

In order to understand roughly how precession changes the Right Ascension and the declination of stars during the 25 800 years cycle the use of the star globe may be found helpful as a model.

Draw on the globe with a fine felt pen the circle showing the positions occupied by the pole at various dates, with centre at the pole of the

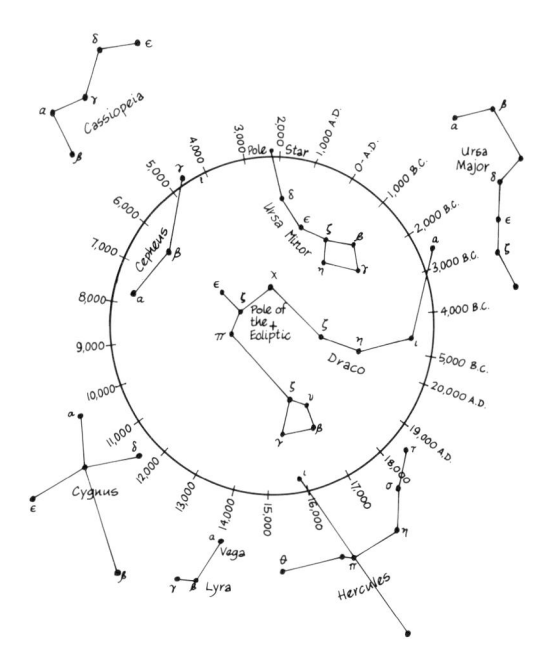

Fig. 9.1. The precession of the pole, showing the positions occupied by the pole at various dates between 5000 BC and 20 000 AD (see Section 9.2).

Reproduced from Sir James Jeans' *The Stars in Their Courses*. Cambridge University Press.

ecliptic and radius representing 23.5°, as shown in Fig. 9.1. Using pieces of cotton and a harmless adhesive, join:

(1) The pole for a particular date (say 4000 BC) to the equinox ♈;
(2) The pole to the star.

The angle formed by the two pieces of cotton at the pole is the new Right Ascension. The new declination of the star can also be shown, and estimated.

By repeating these observations using a full range of pole positions spread over 26 000 years at intervals of a few thousand years and plotting the results for a particular star, a rough idea of the way in which the Right Ascensions and declinations of the star will change or has changed over the centuries will be given. These changes can be calculated as shown in Section 9.4.

9.3 Apart from the long period precession of 25 800 years there is a smaller and short period motion of the pole that is somewhat irregular, but has an amplitude of about 9 seconds and a period of about 18.6 years. The Earth's pole in fact describes a wavy curve which threads its way in and out of the main precessional curve.

The cause of this nutation is the fact that the Moon does not move exactly in the ecliptic. The Moon's orbit is inclined to the ecliptic at an angle of approximately 5° (see Section 3.16).

In addition to this there is a motion known as *planetary precession* produced by the gravitational action of the planets on the Earth and which is responsible for a precession of the equinox of about 12 seconds per century and a decrease in the obliquity of the ecliptic of about 47 seconds a century. These are refinements worth knowing about but their detailed study is outside the scope of this book (see Section 9.5).

9.4 The two systems of coordinates referred to in Section 9.2 have a special point in common, namely the intersection of the plane of the ecliptic and the equator represented as we have seen earlier by the symbol ♈. This point is 90° from the celestial pole along the great circle (RA) = 0 and also 90° from the ecliptic pole along the longitude = 0.

A procedure for finding the approximate declination and RA of a star say, N years ago is as follows:

(1) Transform a star's present equatorial coordinates, declination and RA, to the corresponding ecliptic coordinates latitude and longitude by solving a spherical triangle, see Fig. 9.2, i.e., we find the present latitude and longitude of the star, using a calculator.

Precession of the Earth's Axis

(2) From the latitude and longitude of the star on this present date we can calculate what these coordinates were N years ago by decreasing the longitude by $50'' \times N$ but keeping the *latitude* the same (as the latitude of a star for this exercise can be regarded as constant).

(3) Finally we transform the calculated longitude and latitude for N years ago to RA and declination coordinates and so find what these were at that past date.

This simple treatment, based on that given by Hogben in *Science for the Citizen*, should be regarded as an exercise in the use of the calculator in dealing with another set of spherical triangles which govern the transformation of star coordinates from a reference system based in the plane of the celestial equator, to a system referred to the plane of the ecliptic. For a rigorous treatment of precession more advanced works on positional astronomy should be consulted.

Consider Fig. 9.2. S is a star, declination δ, referred to the plane of the equator, and having a Right Ascension ΥPS. P' is the pole of the ecliptic, P is the pole of the Earth. $P'PS$ is a spherical triangle in which $\delta = PQ - PS = 90 - PS$. $PS = (90-\delta)$, and celestial latitude $ST = (90-P'S)$. Longitude $= 90 - PP'S$ and $P'PS = (90 + (\text{RA}))$.

Apply the cosine formula to find the latitude, which does not change with time.

$$\cos P'S = \cos P'P \cos PS + \sin P'P \sin PS \cos((\text{RA})+90). \quad [9.4(1)]$$

This can be expressed as:

$$\sin(\text{Lat}) = \cos 23\tfrac{1}{2} \sin \delta - \sin 23\tfrac{1}{2} \cos \delta \sin (\text{RA}). \quad [9.4(2)]$$

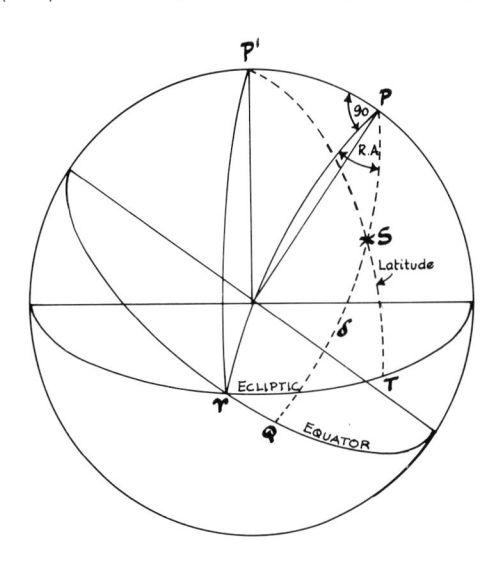

Fig. 9.2

Also, from the spherical triangle $P'PS$ in Fig. 9.2 we have from the cosine formula of Fig. 3.11:

$$\cos PS = \cos P'P \cos P'S + \sin P'P \sin P'S \cos PP'S.$$

This is equivalent to:

$$\sin \delta = \cos 23°.44 \sin(\text{Lat}) + \sin 23°.44 \cos(\text{Lat}) \sin(\text{Long}). \quad [9.4(3)]$$

9.5 Apply to an Actual Star

We can use the two relations [9.4(2)] and [9.4(3)] to find, for example, the coordinates of the declination and the RA of Sirius at 2000 BC. First calculate the present latitude of Sirius referred to the ecliptic using relation [9.4(2)] above.

$$\sin(\text{Lat}) = \cos 23°.44 \sin -16°.68 - \sin 23°.44 \cos 16°.68 \sin 101°$$

whence $\qquad \text{Latitude} = -39°.597.$

We use this latitude to find the present longitude.

The present longitude, 1977, is given by the sine formula of Section 3.4, i.e.

$$\cos(\text{Long}) = \frac{\cos(\text{RA}) \cos \delta}{\cos(\text{Lat})}$$

$$= \frac{\cos 16.68 \cos 101}{\cos 39.597}.$$

The calculator gives longitude $= 103°.72$.

To *this* longitude we apply the shift for precession at the rate of 50″ for each year since 2000 BC, $= (3977 \times 50″) = 55°24.$, which makes the longitude in 2000 BC as $103.72 - 55.24 = 48°.5$ approx.

To find the declination, δ_1, in the year 2000 BC we then use [9.4(3)].

$$\sin \delta_1 = \cos 23.44 \sin -39.597 + \sin 23.44 \cos -39.597 \sin 48.5$$

whence $\qquad \delta_1 = -20°.807$

so that the declination of Sirius in 2000 BC was $-20°.8$.

This method of calculating approximately a star's position several thousands of years ago assumes that the star's celestial latitude is constant and ignores the small changes in the inclination of the Earth's axis to the plane of the ecliptic mentioned in Section 9.3.

These small changes can give rise to errors of three or four degrees when

Precession of the Earth's Axis

the time interval extends to several thousand years for those stars unfavourably situated for this treatment.

Our calculation of the declination of Sirius in the year 2000 BC was $-20°.8$, whereas the calculation taking all factors into account gives the declination as $-19°.40$ in *this* year, see Table 9.1. The calculation for Crucis β however agrees to within $0.1°$ with the published tables.

9.6   Here is another exercise for the calculator, following the same procedure as for Sirius.

It is a little surprising to learn that the Southern Cross was visible from the latitude of New York (41°) about 2000 years ago, and that 4000 years ago it was visible from the south of England (lat 51°).

We can check this by finding the declination of one of the stars in the Southern Cross, for example Crucis β, in the year 2000 BC.

Today Crucis β has a declination of $-59°.32$ and an RA $191°.525$ ($12^h 46^m.1$). As was done in the case of Sirius in Section 9.5, we first find this position in terms of celestial latitude and longitude with reference to the plane of the ecliptic using the spherical triangle transformation formulae [9.4(2)] and [9.4(3)].

$\sin(\text{Lat}) = \cos 23.44 \sin \delta - \sin 23.44 \cos \delta \sin(\text{RA})$

$= \cos 23.44 \sin(-59°.32) - (\sin 23°.44 \cos -59°.32 \sin 191°.525)$

whence $\qquad\qquad\qquad \text{Latitude} = -48°.46.$

Use this latitude to find the present longitude.

$$\cos(\text{Long}) = \frac{\cos \delta \cos(\text{RA})}{\cos(\text{Lat})}$$

$$\cos(\text{Long}) = \frac{\cos(-59°.32) \cos 191°.525}{\cos 48°.46}$$

Longitude $= 221°.069.$

Now using shift in longitude $-50''.25$ per year, the longitude 4000 years ago was

$$221.069 - \left(50.25 \times \frac{4000}{3600}\right) = 221°.069 - 55.83 = 165°.24.$$

Now use the present latitude and the longitude of 4000 years ago to find the declination at that time (δ_1).

Using [9.4(3)],

$$\sin \delta_1 = \cos 23°.44 \sin(\text{Lat}) + \sin 23°.44 \cos(\text{Lat}) \sin(\text{Long})$$
$$\sin \delta_1 = \cos 23°.44 \sin -48°.46 + \sin 23°.44 \cos 48°.46 \sin 165°.26$$

whence

$$\text{declination} = -38°.28$$

For RA

$$\cos(\text{RA}) = \frac{\cos(\text{Long}) \cos(\text{Lat})}{\cos \delta} = \frac{\cos 165°.26 \cos 48°.46}{\cos 38°.28}$$

$$(\text{RA}) = 144°.78.$$

The coordinates of Crucis β were, in 2000 BC, approximately

$$\text{declination } -38°.28 \quad \text{RA } 145°$$

so that Crucis β would be 1° above the horizon in lat 51° (90−51 = 39).

This result agrees with that given in the 5000 and 10 000 year star catalogues published by the Smithsonian Institution, Washington D.C., 1967.

9.7 A few calculations will show that the declinations and Right Ascensions of stars change during the course of a few thousand years in a manner that can be roughly estimated by means of a star globe with its axis made to precess round the pole of the ecliptic, as explained in Section 9.2.

Table 9.1 shows how the Right Ascension and the declinations of three widely separated stars have changed between 2500 BC and the present day, with figures for the year 2500.*

It will be noticed that while Sirius has changed its declination by about 3° during the past 4000 years, Crucis β has changed its declination by about 26°.

The 'Pole Star' (Ursa Minor α) has only in the last 500 years or so been a good navigation aid, as the North Pole Star, and will be nearest in declination to the celestial pole about AD 2100 and thereafter its declina-

---

*The figures are taken, with acknowledgements, from the 5000 year star catalogue published by the Smithsonian Institution, Washington D.C. in Volume 10, No. 2 of the *Smithsonian Contribution to Astrophysics*.

Precession of the Earth's Axis

tion will become less, and about the year AD 4000 it will be 20° or so away from the celestial pole.

It has been suggested that the approach of the star Alpha Ursa Minor (Polaris) to the celestial pole played an important part in the development of navigation from about the 12th Century.

Figures 9.3, 9.4 and 9.5 illustrate how RA and declination of the stars Crucis β, Polaris and Sirius have changed with time.

Table 9.1

Year	Polaris RA (°)	Polaris Declination (°)	Sirius RA (°)	Sirius Declination (°)	Crucis β RA (°)	Crucis β Declination (°)
−2500	331.47	65.05	52.03	−20.85	140.76	−36.08
−2000	334.82	67.57	57.43	−19.40	145.71	−38.35
−1500	338.18	70.14	62.85	−18.17	150.75	−40.75
−1000	341.59	72.81	68.30	−17.19	155.88	−43.27
−500	345.07	75.50	73.77	−16.45	161.16	−45.89
0	348.68	78.24	79.25	−15.97	166.63	−48.59
1000	357.07	83.79	90.27	−15.81	178.40	−54.14
1500	3.76	86.58	95.78	−16.13	148.88	−56.93
2000	37.96	89.26	101.29	−16.72	191.93	−59.69
2500	170.80	87.72	106.79	−17.56	199.70	−62.36

Positional Astronomy and Astro-Navigation Made Easy

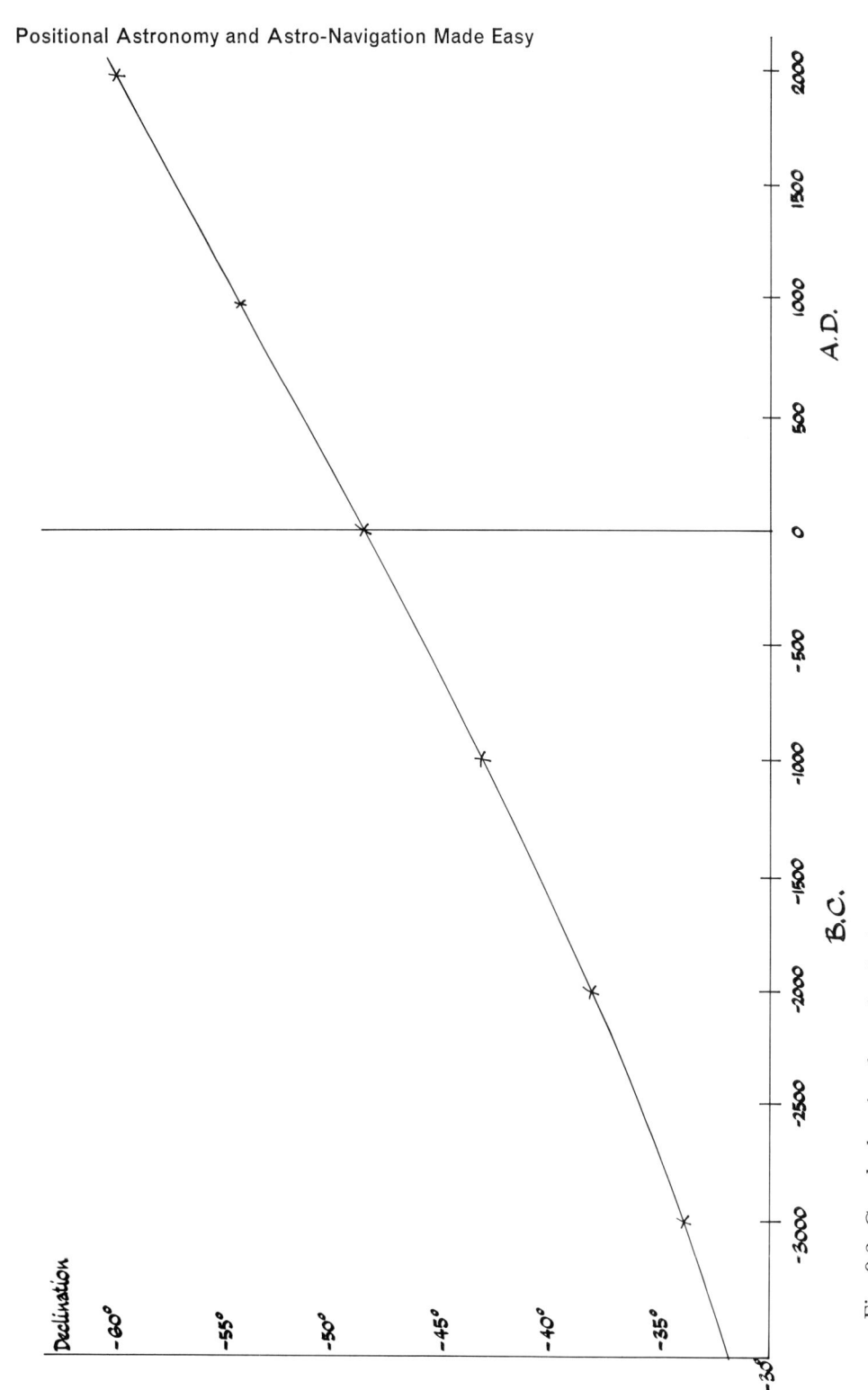

Fig. 9.3. Graph showing how the declination of the star Crucis β in the Southern Cross has changed over the past five millenia. Star visible from lat 41°N, about 100 BC and visible from the south of England, lat 51°N, about 2000 BC.

Precession of the Earth's Axis

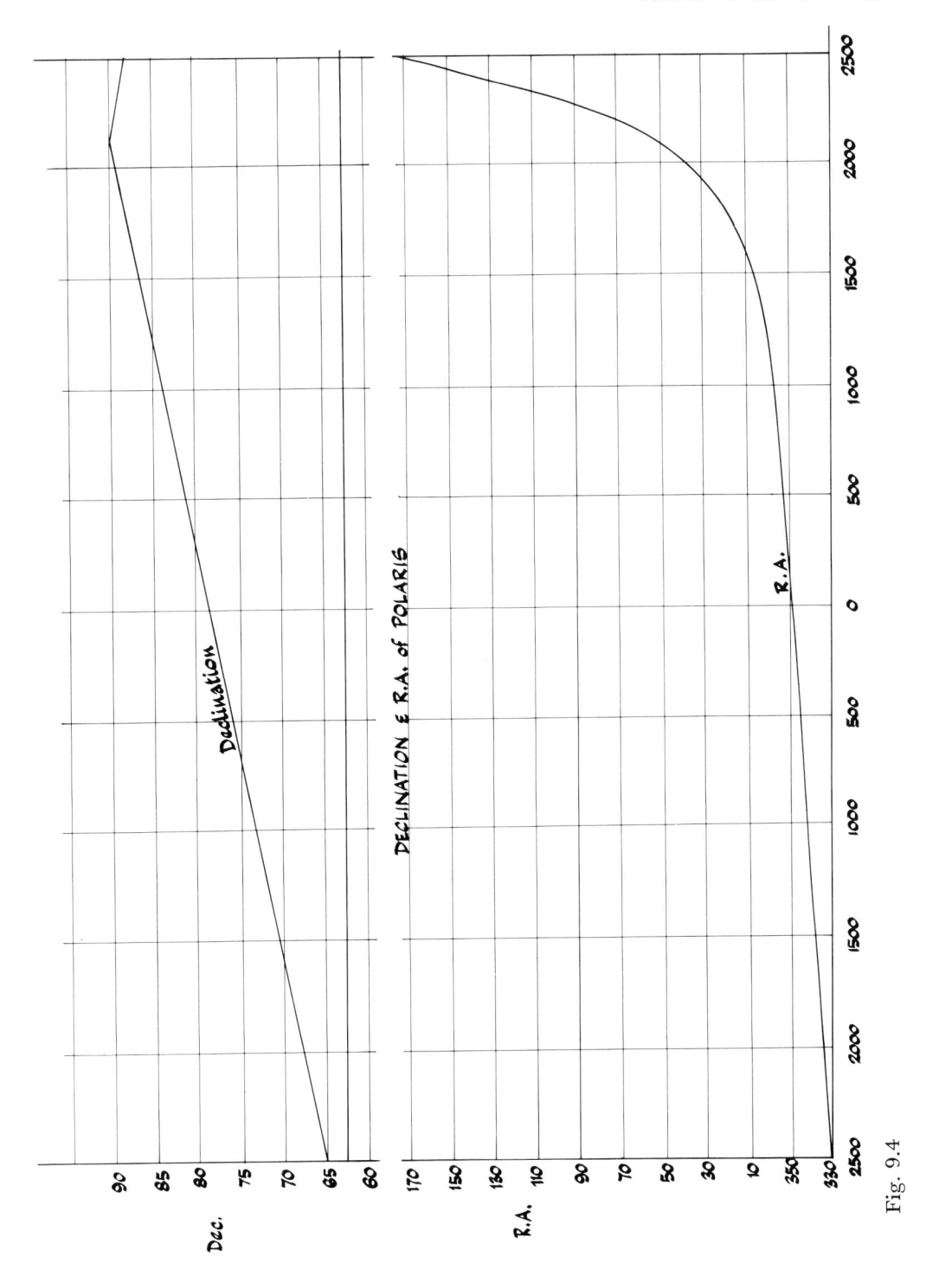

Fig. 9.4

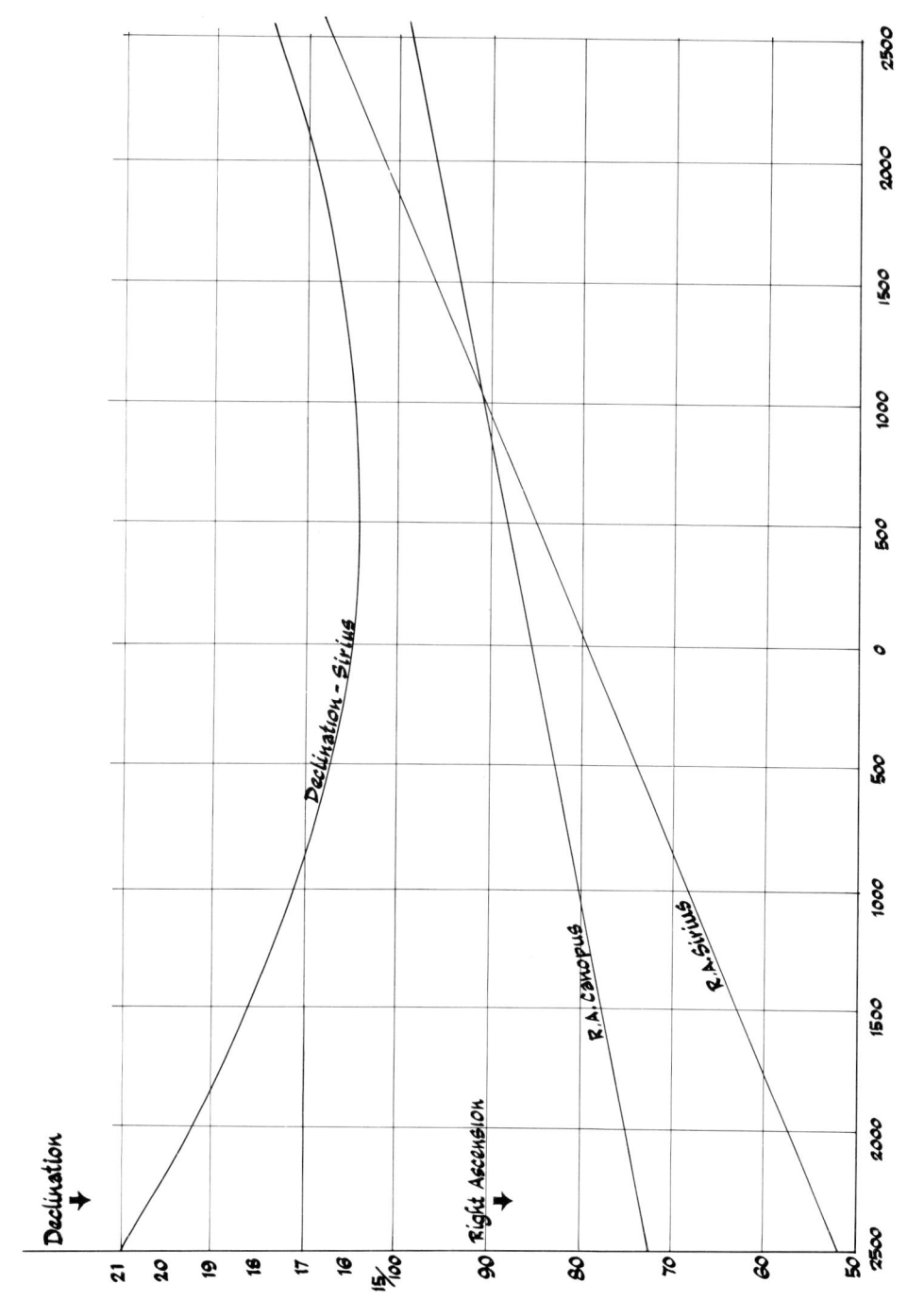

Fig. 9.5

10
Miscellaneous Calculations

How Altitudes and Azimuths Change with Hour Angle

Stellar Magnitudes—Made Easy to Calculate

Simple Calculations Concerning Satellites in Orbit

Positional Astronomy and Astro-Navigation Made Easy

10.1 How Altitudes and Azimuths Change with Hour Angle

Most amateur astronomers and astro-navigators are interested to know how altitudes and azimuths change with time or with declination. The relation connecting altitudes with the hour angle for a celestial body for a particular latitude ϕ and declination δ is,

$$\sin(\text{alt}) = \sin\phi \sin\delta + \cos\phi \cos\delta \cos(\text{HA}).$$

One of the questions that arises in positional astronomy and in navigation is, how does a change in the hour angle of 1° affect the altitude of a star?

10.2

In the relation above the simplest and obvious way to answer the question is to work it out on the calculator. The change in altitude will depend on ϕ, δ, and on HA. Suppose $\phi = 51$, $\delta = 17$ and $(\text{HA}) = 37$. Then,

$$\sin(\text{alt}) = \sin 51° \sin 17° + \cos 51° \cos 17° \cos 37°$$

$$(\text{alt}) = 45°.060\ 413\ 6.$$

Now *increase* HA by 0°.5 to 37°.5, then $(\text{alt}) = 44°.803\ 136\ 1$.

Thus for an increase in hour angle of 0°.5 the altitude falls by 0°.257 277. We can say that 0°.5 of hour angle changes the altitude by 15'.436 62. Suppose the altitude is to be *decreased* by 0°.5 to 36°.5. The altitude would then be 45.315 871 8.

$$\text{the difference} = 0°.255\ 458\ 3 = 15'.327\ 492.$$

We can state that the rate of change of the altitude with respect to a change in hour angle at the altitude of 37° is 30'.764 per degree of altitude or 0°.512 73.

In terms of the calculus, this is expressed as,

$$\frac{d(\text{alt})}{d(\text{HA})} = 0.512\ 7 \text{ for the problem posed above.}$$

10.3

Some may prefer the use of elementary calculus. Then,

$$\sin(\text{alt}) = \sin\phi \sin\delta + \cos\phi \cos\delta \cos(\text{HA}).$$

Differentiate:

$$\cos(\text{alt})\ d(\text{alt}) = \cos\phi \cos\delta \sin(\text{HA})\ d(\text{HA}) \quad (\sin\phi \sin\delta \text{ is constant}).$$

Miscellaneous Calculations

Therefore, $\dfrac{d(\text{alt})}{d(\text{HA})} = \dfrac{\cos \phi \cos \delta \sin(\text{HA})}{\cos(\text{alt})} = \dfrac{\cos 51° \cos 17° \sin 37°}{\cos 45°.0604}$

$= 0.512\ 748\ 559$

$= 0.5127$

which is the same ratio as we worked out in Section 10.2.

10.4

We can use this simple relation of Section 10.3

$$\dfrac{d(\text{alt})}{d(\text{HA})} = \dfrac{\cos \phi \cos \delta \sin(\text{HA})}{\cos(\text{alt})}$$

to find how altitudes are affected by small changes in hour angles (i.e., in time) or vice versa, and we can use it to find out what the effect is of a small error in our hour angle, e.g., through a faulty clock, or a small error in longitude. The problem is related to the common astro-navigation problem of finding the effect on the position line of an error of, say, 10 seconds in time of observation.

For example, what effect on the calculated altitude H_c of Section 4.2 will a clock error of 10 seconds make in the position line under the conditions of the problem of Section 10.2?

$$\dfrac{d(\text{alt})}{d(\text{HA})} = 0.5127$$

but d(HA) is 10 seconds and this is equivalent to 2'.5 of arc. Therefore,

$$\dfrac{d(\text{alt})}{2.5} = 0.5127$$

$$d(\text{alt}) = 2.5 \times 0.5127 = 1'.281\ 75$$

which is just over $1\frac{1}{4}$ nautical miles, and represents the error in the position line arising from an error of 10 seconds in time, under the given conditions.

It is interesting to note that when the hour angle is zero, the rate of change of altitude with respect to hour angle is also zero. At meridian transit the altitude is $(90-51+17) = 56°$ and when the hour angle is small the effect of a small error in time is negligible, as the following example will show.

Suppose the hour angle to be 2°. Then the altitude is given by

$\sin(\text{alt}) = \sin 51° \sin 17° + \cos 51° \cos 17° \cos 2°.$

$\sin(\text{alt}) = 55°.962\ 45$

which is near the transit altitude of 56° given by $(90-\phi+\delta)$.

Now 10 seconds in time is equivalent to $0°.041\ 67$ in hour angle, the altitude is therefore given by

Positional Astronomy and Astro-Navigation Made Easy

$$\sin(\text{alt}) = \sin 51° \sin 17° + \cos 51° \cos 17° \times \cos 2°.041\ 67.$$

Then 10 second error in the clock will make the hour angle 2°.041 67 and the altitude then becomes 55°.960 874 3, and difference of 0°.001 575 7 = 0′.0945 from that given by the correct hour angle.

Thus the effect of a 10 second error in time when taking the altitude of a body that is within 2° of its meridian passage, is (in lat 51° and with 17°) less than 0.1 nautical miles.

The harvest moon. We saw in 3.9 that a celestial body's hour angle at rising or setting was given by $\cos(\text{HA}) = -\tan\phi \tan\delta$. This relation can be applied to the Moon to explain the harvest moon or the early rising of the full moon and the prolonged moonlight about the time of the autumnal equinox. At this time the daily retardation of the Moon, an average of about 50 minutes per day, is considerably reduced by about 20 minutes.

The full moon at this time has a declination of approximately 0° but this declination is changing more rapidly daily than at other times. This can be verified by referring to an astronomical year book. It is about 4.2° in 24 hours.

To find how the HA at rising changes with the Moon's rapidly changing declination, we find by differentiating $\cos(\text{HA}) = -\tan\phi \tan\delta$ that

$$\sin(\text{HA})\,\text{d}(\text{HA}) = \tan\phi\,\text{d}\delta/\cos^2\delta$$

At the autumnal equinox at full moon (HA) = 90° and $\delta = 0°$ approximately so that $\sin(\text{HA}) = 1$ and $\cos\delta = 0$

therefore $\qquad\qquad \text{d}(\text{HA}) = \tan\phi\,\text{d}\delta \qquad\qquad$ (put $\phi = 51°$N).

To obtain the effect of a 4.2° change in declination we have,

$$\text{d}(\text{HA}) = 1.2349 \times 4.2°$$
$$= 5.18657°$$
$$= 5.18657 \times 4 \text{ minutes}$$
$$= 20.75 \text{ minutes}.$$

A farmer at harvest time at full moon has this much more full moonlight for gathering in his harvest.

The graph in Fig. 3.15 shows the effect of a change in declination of 4° on the hour angle at rising.

10.5 Using the methods of Section 10.3 we can find the effect of a small change in hour angle or time, on the *azimuth* of a body. From the relation,

$$\sin(\text{Az}) = \frac{\sin(\text{HA})\cos\delta}{\cos(\text{alt})}$$

Miscellaneous Calculations

we obtain by differentiation, keeping the declination and altitude constant,

$$\cos(\text{Az}) \, d(\text{Az}) = \frac{\cos(\text{HA}) \, d(\text{HA}) \cos \delta}{\cos(\text{alt})}$$

$$\frac{d(\text{Az})}{d(\text{HA})} = \frac{\cos(\text{HA}) \cos \delta}{\cos(\text{Az}) \cos(\text{alt})}.$$

We can apply this in latitude 60° to the star Capella (declination 46). In this latitude Capella is a circumpolar star and has two meridian altitudes, one to the north of the Pole and the other to the south. The rates of change of azimuth with hour angle can be compared in the two cases.

(1) Upper transit, altitude $= (90-60+46) = 76°$

$$\frac{d(\text{Az})}{d(\text{HA})} = \frac{\cos(\text{HA}) \cos 46}{\cos(\text{Az}) \cos(\text{alt})}.$$

When on the meridian $(\text{HA}) = 0$ and $(\text{Az}) = 0$ or 180.

Then, $\quad \dfrac{d(\text{Az})}{d(\text{HA})} = \dfrac{\cos 46}{\cos 76} = 2.8714.$

For a change in hour angle of 1° or 4 minutes in time the azimuth changes 2°.8714. That is, the rate of change is 0'.717 85 per minute of time, or 43' per minute of time.

(2) At the lower transit, the altitude will be 16°, and

$$\frac{d(\text{Az})}{d(\text{HA})} = \frac{\cos 46}{\cos 16} = 0°.722 \, 65 \text{ in 4 minutes}$$

or 10'.84 per minute.

The rate of change of azimuth with HA at upper transit is 4 times that at lower transit.

10.6 The calculation carried out in Section 10.5 will show very clearly the mistake of using a 'watch compass' which when horizontal indicates only *azimuths*, whereas the Sun moves uniformly with the *hour angle* but equatorially (see Section 5.15).

In the summer in latitude 51° and when the Sun's declination is about +23°, the rate of change of azimuth of the Sun at noon with respect to the hour angle or Sun's time is

$$\frac{d(\text{Az})}{d(\text{HA})} = \frac{\cos 23}{\cos 62} = 1.96$$

so the Sun is 'scoring' about 2° of azimuth to 1° of hour angle, and the 'watch compass' of Section 5.9 is sadly out.

In the winter, $\delta = 23°$, the rate of change of azimuth with respect to hour angle is

$$\frac{\cos 23}{\cos 16} = 0.9576.$$

In the winter therefore, at noon, watch and Sun's azimuth run fairly close together (see Table 5.12).

10.7 Simple Calculations Concerning Satellites in Orbit

So far we have considered the calculator as a means of making trigonometrical calculations easy to carry out, but most scientific calculators have several additional function keys that greatly facilitate work in logarithms, natural and common, and problems involving fractional indices, and very large or very small numbers.

For example, some almanacs or astronomical year books give a table showing a list of satellites orbiting the Earth at various heights from the surface of the Earth. Their respective periods (times for a complete orbit) and velocities in kilometres per hour are also given. These are of special interest to satellite spotters.

The relations connecting these quantities are easily deduced from Newton's law of gravitation but involve fractional indices.

Consider a satellite of mass, m in circular orbit round the Earth with a velocity, v, at a distance from the centre of the Earth of R. The mass of the Earth is M.

The acceleration of the satellite towards the centre of the earth is v^2/R, and this is caused by the gravitational attraction of the Earth which is given by the universal law of gravity

$$F = \frac{GmM}{R^2}$$

and this $= m \times$ acceleration towards the centre. We have therefore,

$$\frac{mv^2}{R} = \frac{GmM}{R^2} \qquad [10.7(1)]$$

where G is the gravitational constant.

If the periodic time of the orbit is T, then $vT = 2\pi R$

$$\text{or} \quad T = \frac{2\pi R}{v}$$

$$\text{or} \quad v = \frac{2\pi R}{T}.$$

Substituting the value of v in equation [10.7(1)] we have,

$$\frac{m 4\pi^2 R^2}{T^2 R} = \frac{GmM}{R^2}$$

or,
$$T^2 = \frac{(2\pi)^2 R^3}{GM}$$

$$T = 2\pi \sqrt{\frac{R^3}{GM}} \qquad [10.7(2)]$$

It is convenient to consider R as being equal to $r+h$ where r is the radius of the Earth and h the height of the satellite above the Earth. Therefore,

$$T = 2\pi \sqrt{\frac{(r+h)^3}{GM}}.$$

A practical problem of some interest suggests itself, namely, at what height above the Earth should a satellite be orbiting in order that the time of its orbit would be 24 hours? What is the value of h for a geosynchronous orbit, which is now used in television systems, when an apparently permanent satellite over a particular region is required?

We have from equation [10.7(2)],

$$R^3 = \frac{T^2 M}{4\pi^2}.$$

where $T = 24 \times 60 \times 60$ sec, $G = 6.670 \times 10^{-11}$ kg^{-1} m^3 s^{-2}, $M = 5.98 \times 10^{24}$ kg.

$$R^3 = \frac{(24 \times 3600)^2 \times 6.670 \times 10^{-11} \times 5.98 \times 10^{24}}{4\pi^2}$$

$$= \frac{(2.4 \times 3.6 \times 10^4)^2 \times 6.670 \times 10^{-11} \times 5.98 \times 10^{24}}{4\pi^2}$$

$$R^3 = \frac{2977.52}{4\pi^2} \times 10^{21}$$

$$= 75.42 \times 10^{21} \text{ m}$$

$\therefore \quad R = 4.225 \times 10^7$ m.

The radius of the Earth, r, is 6.378×10^6 m.

But $\quad R = r+h = 4.225 \times 10^7 = 42.25 \times 10^6$ m

but subtracting the radius of the Earth $r = 6.38 \times 10^6$ m

$\therefore \quad h = (R-r) = 35.87 \times 10^6$ m

Therefore h (height above surface of Earth) is 35.87×10^6 m or, $h = 35\,870$ kilometres which is the recognised approximate position for a geosynchronous satellite.

Positional Astronomy and Astro-Navigation Made Easy

10.8 Stellar Magnitudes—Made Easy to Calculate

Hipparchus 2000 years ago made a classification of the stars visible to the naked eye according to their apparent brightness. The fifteen brightest stars were put at the top of the class and were designated 'Stars of the 1st Magnitude'. Stars just visible were called 'Stars of the 6th Magnitude'. So magnitude is simply a number associated with the brightness of a star.

The main function of an astronomical telescope is not so much to magnify objects but to bring a great deal more light to the eye than the eye can collect unaided.

The important question to ask about a telescope is, "What is the diameter of its object glass or mirror?"

This information tells us what stars we can see which are beyond the power of our unaided eyesight. The amount of light that can be gathered by an object lens of D mm is proportional to D^2. The amount of light that the unaided eye can gather is proportional to d^2 where d is the diameter of the entrance pupil of the eye adjusted for darkness; this is about 8 mm.

The light-gathering power (or advantage) of the telescope, assuming all the light entering it is brought into the eye, is D^2/d^2.

For a modest telescope of 75 mm diameter and for a normal eye the telescope brings in $75^2/8^2$ more light than can be received by the eye, $= 87.89$.

The perceptiveness of our eyes does not follow the same equal steps which we would expect from physical optics. For example, if a star, A, sends us just 100 times more light than another, B, when B is just visible, we would judge that star B to be five steps brighter than A, i.e., about as bright as Aldebaran or Altair which have a magnitude of 1. If the size of each step is x then, $x^5 = 100$ and therefore $x = 2.512$. The scale of magnitudes is such that a star of one magnitude brighter than another is therefore 2.512 times as bright.

In a 75 mm telescope which increases the amount of light from a star by a factor of 87.89, we can say that it changes the magnitude of a star when viewed through the telescope by x, where

$$2.512^x = 87.89$$

$$x = \frac{\log 87.89}{\log 2.512} = 4.86 \quad \text{(using the calculator)}.$$

So a star barely visible to the naked eye (magnitude 6) would appear as a star of magnitude $6 - 4.86 =$ about 1, or comparable with Altair when viewed through the telescope.

Miscellaneous Calculations

You may wonder how it is possible to measure the altitude of the Pole Star (magnitude 2) in the nautical twilight when the star is barely visible to the naked eye. Sextants are fitted with small telescopes having an object lens of about 28 mm. The eye has, say, a pupil diameter of 6 mm. Then the sextant telescope produces a light gain of $28^2/6^2 = 21$. This improves the appearance of the star Polaris by x magnitudes, where as before, $2.512^x = 21$ and the calculator gives $x = 3.3$. So Polaris (magnitude 2) would appear to the navigator in his sextant telescope as magnitude $(2-3.3) = -1.3$, very nearly as bright as Sirius, so that measuring its altitude even in twilight would be no problem.

Lists of stars are given in *Norton's Star Atlas*, showing for each star its distance from our system in 'light years'. The apparent magnitude of the star and the star's absolute magnitude, i.e., as it would appear at 33 light years, or at a distance of 10 parsecs (see Appendix III).

An instructive exercise can be carried out using the calculator to check the results given. For example, Antares, which is at a distance of ≈ 400 LY and has an apparent magnitude of 0.92 would appear $(400/33)^2$ times as bright at 33 LY distance, or 146.92 times as bright.

If x is the number of magnitudes *brighter*, then

$$2.512^x = 146.92$$
$$x = 5.417 \text{ magnitudes.}$$

Antares absolute magnitude $= 0.92 - 5.417 = -4.5$, as shown in star tables.

10.9 A question the amateur astronomer often asks is what is the limiting magnitude, M, of stars that can just be seen in a telescope having an objective of D mm, assuming that a star of magnitude 6.5 can just be seen by the naked eye, assumed to have a pupil diameter of 8 mm?

If x is the increase in magnitude caused by the increase in light-gathering power measured by the ratio $D^2/8^2$ then

$$2.512^x = \frac{D^2}{8^2}$$

$$2.512^{(M-6.5)} = \frac{D^2}{8^2}$$

$$(M-6.5) \log 2.512 = 2 \log_{10} D - 2 \log 8$$

$$\log_{10} 2.512 = 0.4$$

$$M - 6.5 = 5 \log_{10} D - 4.5$$

Therefore, $$M = 2+5\log_{10} D. \quad [10.9(1)]$$

This is a convenient formula for the calculator but it is reliable only under ideal seeing conditions.

Example: The limiting magnitude of stars visible in a telescope of $D = 300$ mm is
$$M = 2+5\log_{10} 300$$
$$= 14.39.$$

10.10 A useful formula which gives the absolute magnitude of a star in terms of the star's apparent magnitude and the star's distance in parsecs can be derived as follows, and is similar to equation [10.9(1)].

If b is the brightness of a star at distance d and B the brightness of the star at the standard distance, S, then $b/B = S^2/d^2$.

Let M be the absolute magnitude of the star, i.e., at a distance S, which by convention = 10 parsecs, and let m be the apparent magnitude of the star at distance d. Then,
$$2.512^{(M-m)} = \frac{S^2}{d^2}$$

Taking logs, $$(M-m)\log_{10}(2.512) = 2\log_{10}\frac{S}{d} \quad [10.10(1)]$$

The ratio of the distances $S/d = \pi/\pi_0$ where π_0 is the parallax of the star at distance S and π is the parallax at distance d. A star at the standard distance S of 10 parsecs has a parallax of $0''.1$.

So we can write [10.10(1)] above, since $\log 2.512 = 0.4$, as
$$M-m = 5\log_{10}\pi - 5\log_{10}\pi_0$$
but $$\log_{10}\pi_0 = \log_{10} 0.1 = -1$$
$\therefore$ $$M-m = 5\log_{10}\pi + 5.$$

The required formula for the absolute magnitude
$$M = m+5+5\log_{10}\pi.$$

π, the star's parallax, is given in most star tables and $1/\pi$ = distance of a star in parsecs.

Postscript

In this book there are inevitably many omissions and some rather over-simplified explanations of mathematical relations used, but the work may encourage further study and suggest other projects involving positional astronomy. For serious more advanced study, books on spherical astronomy and positional astronomy listed in the Bibliography should be consulted.

The star globe has been mentioned as a valuable visual aid in understanding the celestial sphere but mention should be made here of the modern planetarium and the help it can give to all who wish to study and understand the principles of positional astronomy. Planetaria are now established in many of the large cities of the world and provide not only special sessions in astronomy, as a part of general education, but also introductory courses in astro-navigation. The planetarium in its various forms has played an important role by presenting a working model of the celestial sphere which simulates with remarkable accuracy and clarity the star-filled sky and the apparent motions of all visible celestial bodies.

Planetaria from the days of simple globes and orreries of the 18th Century to the modern sophisticated optical projection devices have thus been effectively used in teaching astronomy and navigation. They help to clarify the coordinate systems of stars on the celestial sphere and to promote a proper understanding of spherical triangles. Once these are fully understood the electronic calculator makes positional astronomy easy.

The book was begun under the title *Astronomical Fun with a Calculator,* and was centred round a few simple projects and problems in positional astronomy and astro-navigation that called for modern calculator methods for numerical solutions. Notes on the subject were prepared for talks to small astronomical groups, but as the number of things to do and to make using a calculator increased, it was felt worthwhile to increase the scope of the title to its present form, but it is hoped that the astronomical fun will persist despite the rather pretentious title. Astronomy is not merely an academic discipline in professional education, but has always been, and still is, a matter for enjoyment for all who wonder at the beauty of our Universe

I conclude in the words of Sir Thomas Browne the eminent Physician and literary figure of the 17th Century (1605–1682).

> "Often have I admired the mystical way of Pythagoras, and the secret magic of numbers. For there is a music wherever there is harmony, order or proportion, and thus we may maintain the music of the spheres, for those well ordered motions and regular paces, though they give no sound unto the ear, yet do they strike a note most full of harmony."

Appendix I

As mentioned in Section 3.12, the relation [3.7(8)]

$$\cot(\text{Az}) = \frac{\cos \phi \tan \delta - \sin \phi \cos(\text{HA})}{\sin(\text{HA})}$$

can be deduced from relations [3.7(1)], [3.7(4)] and [3.7(7)], and as relation [3.7(8)] is of special importance in finding azimuths, *knowing only δ, ϕ and HA*, it is worthwhile going through the elementary trigonometry involved.

Relation [3.7(1)] is

$$\sin(\text{alt}) = \sin \phi \sin \delta + \cos \phi \cos \delta \cos(\text{HA})$$

and is of the same form as [3.7(4)]

$$\sin \delta = \sin \phi (\text{alt}) + \cos \phi \cos(\text{alt}) \cos(\text{Az})$$

Relation [3.7(7)] is

$$\sin(\text{Az}) = \frac{\sin(\text{HA}) \cos \delta}{\cos(\text{alt})}$$

Now relation [3.7(8)] does not contain altitude, so we eliminate it from [3.7(1)], [3.7(4)] and [3.7(7)].

From [3.7(4)]

$$\cos(\text{Az}) = \frac{\sin \delta - \sin \phi \sin(\text{alt})}{\cos \phi \cos(\text{alt})}$$

From [3.7(1)] and [3.7(7)]

$$\cot(\text{Az}) = \frac{\cos(\text{Az})}{\sin(\text{Az})} = \frac{(\sin \delta - \sin \phi \sin(\text{alt})) \cos(\text{alt})}{\cos \phi \cos(\text{alt}) \sin(\text{HA}) \cos \delta}$$

$$= \frac{\sin \delta \cos(\text{alt}) - \sin \phi \sin(\text{alt}) \cos(\text{alt})}{\cos \phi \cos(\text{alt}) \sin(\text{HA}) \cos \delta}$$

$$= \frac{\tan \delta}{\cos \phi \sin(\text{HA})} - \frac{\tan \phi \sin(\text{alt})}{\sin(\text{HA}) \cos \delta}.$$

Now

$$\sin(\text{alt}) = \sin \phi \sin \delta + \cos \phi \cos \delta \cos(\text{HA}).$$

Then,

$$\cot(\text{Az}) = \frac{1}{\sin(\text{HA})}\left[\frac{\tan\delta}{\cos\phi} - \frac{\tan\phi}{\cos\delta}(\sin\phi\sin\delta + \cos\phi\cos\delta\cos(\text{HA}))\right]$$

$$= \frac{1}{\sin(\text{HA})}\left[\frac{\tan\delta}{\cos\phi}(1-\sin^2\phi) - \sin\phi\cos(\text{HA})\right]$$

$$= \frac{1}{\sin(\text{HA})}\left[\tan\delta\cos\phi - \sin\phi\cos(\text{HA})\right].$$

Therefore,

$$\cot(\text{Az}) = \frac{\tan\delta\cos\phi - \sin\phi\cos(\text{HA})}{\sin(\text{HA})}.$$

which is the required relation.

Appendix II

Useful Information

(1) *The Sun*

Solar mass	1.990×10^{30} kg
Solar radius	6.960×10^{8} m
Semi-diameter at mean distance	$15'59''$
Astronomical unit (1 AU)	1.496×10^{11} m

(2) *The Earth*

Earth mass	5.976×10^{24} kg
Polar radius	$6.356\ 775 \times 10^{6}$ m
Equatorial radius	$6.378\ 160 \times 10^{6}$ m

(3) *The Moon*

Lunar mass	7.35×10^{22} kg
Lunar radius	1.738×10^{6} m
Semi-diameter at mean distance	$15'32''.6$ (geocentric)
Mean equatorial horizontal parallax	$3422''.54$
Mean distance from Earth	3.8440×10^{8} m
Inclination of orbit to ecliptic	$5°8'43''$

(4) *Time*

24^h mean solar time	$24^h\ 03^m\ 56^s$ mean sidereal time
24^h mean sidereal time	$23^h\ 56^m\ 04^s.09$ mean solar time
Mean solar day	$24^h\ 03^m\ 56^s.555$
	$1^d.002\ 737\ 91$ mean sidereal time
Mean sidereal day	$23^h\ 56^m\ 4^s.091$ mean solar time

Appendices

(5) *Length of year*

Julian	365.25 days
Tropical (equinox to equinox)	365.242 19 days
Sidereal (fixed star to fixed star)	365.256 36 days
Anomalistic (perihelion to perihelion)	365.259 64 days
Eclipse (lunar node to lunar node)	346.620 03 days

(6) *Length of the month*

Synodic (new moon to new moon)	29.530 59 days
Tropical (equinox to equinox)	27.321 58 days
Sidereal (fixed star to fixed star)	27.321 66 days
Anomalistic (perigee to perigee)	27.554 55 days

(7) *Miscellaneous*

Gravitational constant	6.670×10^{-11} kg^{-1} m^3 s^{-2}
Speed of light	$2.997\,925 \times 10^8$ m s^{-1}
Parsec	206 264.8 AU
	3.0857×10^{16} m
	3.2616 light years
Light year	9.4607×10^{15} m
	6.324×10^4 AU
Nautical mile	1.853 km; 1853 m
Annual general precession	$50''.2564 + 0''.0222\ T$ (T measured in Julian centuries from 1900)
1 radian	$57°.295\,78$
1°	0.017 45 radians

Appendix III

Stellar Distances

$\quad$ 1 parsec is a distance of $\quad$ 3.2616 light years

$\quad$ 10 parsecs is a distance of $\quad$ 32.616 light years

A parsec is not an angle but is that distance, D, of a star which gives rise to a parallax of $1''$.

$$\text{Parallax in radians} = \frac{\text{distance earth to Sun } (r)}{\text{distance of star } (D)}$$

This expressed in seconds of arc $= \frac{r}{D} \times 57.295\,78 \times 60 \times 60''$

If parallax angle is $1''$ then, $\quad 1 = \frac{r}{D} \times 2.062\,65 \times 10^5$

but, $\quad r = 1.496\,00 \times 10^{11}$ m $\quad$ and $\quad D = 1.4960 \times 10^{11} \times 2.062\,65 \times 10^5$ m

$\quad$ 1 light year $= 9.4607 \times 10^{15}$ m

$\therefore \quad D$, or 1 parsec $= \dfrac{1.496 \times 10^{11} \times 2.062\,65 \times 10^5}{9.4607 \times 10^{15}}$

$\qquad\qquad\qquad\quad = 3.2616$ light years

The distance of a star in parsecs is the reciprocal of its parallax in seconds of arc.

If, for example, the parallax of a star is $0.026''$ its distance in parsecs is the reciprocal of 0.026, $\quad 1/0.026 = 38.461$ parsecs.

Appendix IV

To facilitate finding the number of days that have elapsed since a particular date, for the purpose of finding the approximate local sidereal time and thence the local hour angle, for use with Figures 5.7 and 5.8, the following may be useful.

Number of days since 21st September

To 1st October	10
To 1st November	40
To 1st December	71
To 1st January	102
To 1st February	133
To 1st March	161
To 1st April	192
To 1st May	222
To 1st June	253
To 1st July	283
To 1st August	314
To 1st September	345

Appendix V

The Right Ascensions and declinations of some of the brightest stars.

Star Name	Constellation	RA h m	Declination ° ′	Apparent magnitude
Alpheratz	α Andromeda	0 07.2	+28 58	2.1
Schedar #	α Cassiopeiae	0 39.2	+56 25	2.3
Mirach	α Andromedae	1 08.4	+35 30	2.4
Achernar	α Eridani	1 36.9	−57 21	0.6
Polaris #	α Ursae Minoris	2 09.2	+89 10	2.1
Algol	β Persei	3 06.7	+40 52	Var.
Mirfak	α Persei	3 22.7	+49 47	1.9
Aldebaran	α Tauri	4 34.6	+16 28	1.1
Rigel	β Orionis	5 13.4	−8 14	0.3
Capella	α Aurigae	5 15.0	+45 59	0.2
Bellatrix	γ Orionis	5 23.9	+6 20	1.7
Betelgeuse	α Orionis	5 53.9	+7 24	Var.
Mirzam	β Canis Majoris	6 21.7	−17 57	2.0
Canopus	α Carinae	6 23.4	−52 41	−0.9
Sirius	α Canis Majoris	6 44.1	−16 41	−1.6
Castor	α Geminorum	7 33.1	+31 56	1.6
Procyon	α Canis Minoris	7 38.1	+5 17	0.5
Pollux	β Geminorum	7 43.9	+28 05	1.2
Regulus	α Leonis	10 07.1	+12 05	1.3
Merak #	β Ursae Majoris	11 00.5	+56 30	2.4
Dubhe #	α Ursae Majoris	11 02.3	+61 53	1.9
Denebola	β Leonis	11 47.9	+14 42	2.2
	α Crucis†	12 25.3	−62 58	1.0
	γ Crucis†	12 29.9	−56 59	1.6
	β Crucis†	12 46.4	−59 34	1.5
Spica	α Virginis	13 24.0	−11 02	1.2
Hadar	β Centauri†	14 02.2	−60 16	0.9
Arcturus	α Bootis	14 14.6	+19 18	0.2
Rigil Kent	α Centauri†	14 38.0	−60 44	0.1
Alphecca	α Coronae Borealis	15 33.7	+26 47	2.3
Antares	α Scorpii	16 28.0	−26 23	1.2
	α Trianguli Australis†	16 46.2	−68 59	1.9
Rasalhague	α Ophiuchi	17 33.9	+12 35	2.1
Vega	α Lyrae	18 36.2	+38 46	0.1
Altair	α Aquillae	19 49.7	+8 48	0.9
Deneb	α Cygni	20 40.6	+45 12	1.3
Fomalhaut	α Piscis Austrini	22 56.4	−29 45	1.3

Appendices

It can be seen from a study of the star globe fitted with a horizon circle (Fig. 3.14) that some stars can never appear above the horizon in a particular latitude, (ϕ).

In northern latitudes where ϕ is positive, all stars having a declination between $+90°$ and $\phi - 90°$ can appear above the horizon. For example in latitude 51°N all stars having declinations between 90°N and $-39°$S can so appear. The six stars marked † in the table, Crucis α, β and γ, Centauri α and β, and Trianguli Australis α cannot appear to observers in latitude 51°N.

In southern latitudes where ϕ is negative, all stars having a declination between $-90°$S and $90° + \phi$ are visible, i.e. can appear above the horizon. For example in latitude $-35°$ (southern Australia) only those stars with declinations between $-90°$S and 55°N can be observed. Only four of the listed bright stars are thus excluded from an observer in this southern latitude, and are marked # (Schedar, Polaris, Merak and Dubhe).

Appendix VI

The table in Appendix IV can be used to find the time, date and hour at which any particular star will appear on the meridian (hour angle = 0°). We can then estimate the star's position a few hours earlier or later than the meridian passage, for the purpose of knowing where to look.

There are two simple questions often raised by amateur observers and casual star gazers:

(1) What stars are best viewed on a particular date, for example D days from 21st September?

(2) When can I see under favourable conditions a particular star of Right Ascension, (RA), and where do I look?

For simplicity we take the best viewing time as midnight because it is then dark *all* the year round between latitudes 65°N and 65°S, and stars can generally best be seen when they are on or near the meridian. This gives a definite and easily found direction in which to look (due south in the northern hemisphere) and stars of 'visible' declinations are at their maximum altitudes.

The two essential facts to be borne in mind are:

(1) That any particular star will transit the meridian when its RA equals the local sidereal time;

(2) Sidereal time gains on mean sun time or civil time about 4 minutes each day (see Section 2.9) starting from 21st September when local mean time and local sidereal time are equal to within a few minutes (see Section 2.4).

Taking the observing time as midnight, then the local sidereal time on a date D days after 21st September will be

$$\frac{D \times 4}{60} \text{ hours} = \frac{D}{15} \text{ hours}$$

Thus on that date stars having (RA) $= \frac{D}{15}$ will culminate on that day, at that time.

A few examples will illustrate the simple approximate answers to the questions asked.

(1) What stars are in a favourable position for viewing on 22nd February? In other words what stars will be culminating at midnight on this date?

$$D \text{ from Appendix IV} = 133 + 22 = 155 \text{ days}$$

$$\therefore \quad \text{stars of (RA)} \frac{155}{15} = 10.33^{\text{h}} = 10^{\text{h}} \ 20^{\text{m}} \quad \text{will culminate then.}$$

Appendices

The star that will be near the meridian will be one having a RA near $10^h 20^m$. This will help identify the star. A glance at Appendix V will identify the star as Regulus (RA) $10^h 07^m.1$. It will transit at an altitude of $(90-\phi+12)$ in latitude ϕ. The stars Procyon, Castor and Pollux, RA about $7^h 44^m$, will have culminated about two and a half hours before midnight, as can be verified from a planisphere or star globe.

(2) When is a star of (RA) $6^h 44^m$ (i.e. Sirius) well positioned for observation?

The star's meridian passage will be at midnight on date D, see Appendix V.

where
$$\frac{D}{15} = (RA) = 6^h 44^m = 6^h.733$$

$$D = 15 \times 6.733 = 101 = 71 + 30.$$

The date will therefore be *31st December*.

Vega with (RA) $18^h 36^m.2$ will culminate at midnight on date D.

where
$$18^h 36^m.2 = \frac{D}{15}$$

$D = 279$ and the date is 16th June. A truly summer star.

It will be appreciated that this method of reckoning is approximate only as we have used 4 minutes instead of $3^m 56^s.56$.

Appendix VII

An Additional Project on The Summer Triangle

The northern hemisphere night sky is dominated during the winter months by the constellation of Orion and during the summer by the stars of the summer triangle, consisting of Vega, Altair and Deneb. This triangle of stars, VAD, in the figure can form the basis of a calculation project in positional astronomy.

For example, it can be used to study the elements of the spherical triangles formed by these three stars.

From the star's positions in terms of RA's and declinations it is possible to calculate the angular distances between the stars, and the angles at each of the stars formed by the other two. In the figure the angular distances, i.e. VD, VA and DA can be found separately by considering each one in turn as part of a spherical triangle of which the pole, P, is the apex as is done in navigation.

In the spherical triangle PVD, which shows the declinations and RA's of each star $PV = (90-\text{declination of } V)$, $PD = (90-\text{declination of } D)$ and $PA = (90-\text{declination of } A)$ also the angle $VPD = $ difference in RA's between Vega and Deneb.

Hence, using the cosine formula of Section 3.4,

$$\cos VA = \cos PV \cos PA + \sin PV \sin PA \cos VPA$$

$$\cos VA = \sin 38°.46 \sin 8°.8 + \cos 38°.46 \cos 8°.8 \cos 18°.375$$

whence

$$VA = 33°.95$$

Similarly,

$$\cos VD = \sin 38°.46 \sin 45°.2 + \cos 38°.46 \cos 45°.2 \cos 31°.10$$

and

$$VD = 23°.966$$

$$\cos DA = \sin 45°.2 \sin 8°.8 + \cos 45°.2 \cos 8°.8 \cos 12°.725$$

$$DA = 38°.0205$$

We now have calculated each of the three sides of the spherical triangle VAD and can now calculate the angles at V, A and D, using the formula of Fig. 3.11 rearranged.

Appendices

P ☀ Pole

D ☀ DENEB
$\delta_D = 45°12'$
$RA_D = 20^h 40.6^m$

23.966°   64°

V ☀ 82.73°

VEGA
$\delta_V = 38°46'$
$RA_V = 18^h 36.2^m$

38.025

33.95

40.82

A ☀ ALTAIR
$\delta_A = 8°48'$
$RA_A = 19^h 49.7^m$

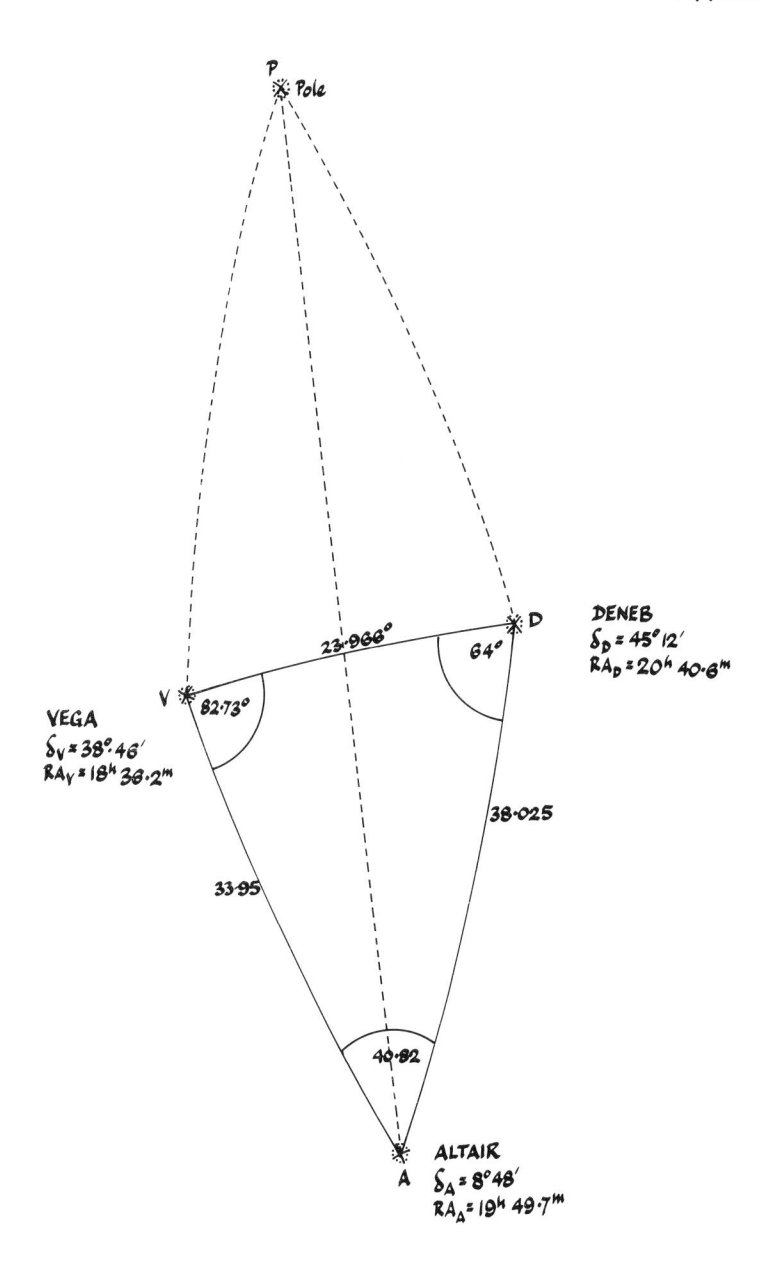

The figure shows the spherical triangle, formed by Vega, Altair and Deneb solved using the RA's and declinations of these stars as a calculator exercise.

261

i.e.
$$\cos V = \frac{\cos 38°.0205 - \cos 23°.966 \cos 33°.95}{\sin 23°.966 \sin 33°.95}$$

whence
$$V = 82°.45$$

$$\cos A = \frac{\cos 23°.966 - \cos 33°.95 \cos 38°.0205}{\sin 33°.95 \sin 38°.0205}$$

$$A = 40°.82$$

$$\cos D = \frac{\cos 33°.95 - \cos 23°.966 \cos 38°.0205}{\sin 23°.966 \sin 38°.0205}$$

$$D = 64°.006$$

It will be noticed that the sum of the three angles $V, A, D = 187°.286$. This should cause no surprise as the sum of the angles of a spherical triangle is always greater than 180°.

The figure gives the solution of the spherical triangle formed by Vega–Altair–Deneb.

It is a good exercise using the cross staff of Section 3.2(3) or the sextant to measure the sides of this impressive triangle of summer stars and compare with the calculated values.

Bibliography

Abbott, P. *Teach Yourself Trigonometry*, English Language Book Society 'Teach Yourself' Books, 1963.
Barlow & Bryan. *Elementary Mathematical Astronomy*, University Tutorial Press, 1944.
Beet, E. A. *Mathematical Astronomy for Amateurs*, David & Co., 1972.
Blewitt, Mary. *Celestial Navigation for Yachtsmen*, Stanford Maritime, 1972.

Chaucer, Geoffrey, edited by Walter Skeet. *A Treatise on the Astrolabe, 1391*, addressed to his son Lowys, 1969.
Chichester, Francis. *The Observer's Book on Astro-Navigation*, Parts I–III, George Allen & Unwin, London, 1941.
Cotter, C. H. *The Elements of Navigation*, Pitman, 1953.
Cousins, Frank. *Sundials*, John Baker, London, 1969.

Evans, David. *Teach Yourself Astronomy*, Hodder & Stoughton, 1975.

Hogben, Lancelot. *Science for the Citizen*, George Allen & Unwin, 1956.

Mayer, Tobias. *The Birth of Navigational Science*, National Maritime Museum Maritime Monograph & Reports No. 10, 1974.
McNally, D. *Positional Astronomy*, Muller Educational, Frederick Muller Ltd., London NW2 6LE, 1974.
Mitton, Simon. *The Cambridge Encyclopaedia of Astronomy*, Jonathan Cape, 30 Bedford Square, London, 1977.
Moore, Patrick. *Astronomy for 'O' Level*, G. Duckworth & Co., 1973.
———. *The Observer's Book of Astronomy*, F. Warne & Co. Ltd., 1971.
———. *Year Book of Astronomy*, Sidgwick & Jackson, annually.

National Maritime Museum, Greenwich London. *The Planispheric Astrolabe*, H.M. Stationery Office, 49 Holborn, London WC1, 1976.
North, John D. *The Astrolabe*, Scientific American, New York, N.Y., U.S.A., January 1974.
Norton, A. P. and Inglis, H. Gall. *Norton's Star Atlas and Reference Handbook*, Gall & Inglis (sixteenth ed.), 1973.

Ronan, C. A. *Astronomy*, David and Charles, Newton Abbot, Devon, 1973.

Saunders, Harold N. *The Astrolabe*, Micro Instruments (Oxford) Ltd., Little Clarendon Street, Oxford, OX1 2HP, 1971.
Schroeder, *Practical Astronomy*, Werner & Laurie.
Sidgwick, J. B. *Amateur Astronomer's Handbook*, Faber & Faber, London, 1971.
Smart, W. M. *Foundations of Astronomy*, Longman, 1944.
———. *Textbook on Spherical Trigonometry*, Cambridge University Press, 1962.

Wallenquist, A. *Dictionary of Astronomy*, Penguin, 1968.
Waugh, Albert E. *Sundials, Their Theory and Construction*, Dover Publications Inc., New York, N.Y., U.S.A., 1973.
Weatherlake, J. G. P. *The EBBCO Sextant*, East Berks Boat Co., Wargrave, Berkshire, 1974.
Webster, Roderick S. *The Astrolabe. Some Notes on its Construction and Use*, Paul MacAlister & Associates, Lake Bluff, Illinois, U.S.A., 1974.

Admiralty Navigation Manual, Volumes I–III, H.M. Stationery Office.
Navigation in the Days of Captain Cook, National Maritime Museum, Greenwich, London SE10 9NF.
Nuffield Physics Background Book—Astronomy, Longman Penguin, 1972.

ALMANACS AND EPHEMERIDES

The American Ephemeris and The Nautical Almanac, Her Majesty's Nautical Almanac Office, London, and the Nautical Almanac Office, United States Naval Observatory, Washington, U.S.A., published annually.
The Astronomical Ephemeris, Her Majesty's Nautical Almanac Office, London, and the Nautical Almanac Office, United States Naval Observatory, Washington, U.S.A., published annually.
Explanatory Supplement to these Almanacs and Ephemerides, prepared jointly by the Nautical Almanac Offices of the United Kingdom and the United States of America, H.M.S.O., London.
Whitaker's Almanack, William Clowes & Sons Ltd., London.

JOURNALS AND PERIODICALS

Hermes, monthly, The Junior Astronomical Society (Secretary), 58 Vaughan Gardens, Ilford, Essex.
Journal of the British Astronomical Association, bi-monthly, The British Astronomical Association, Burlington House, Piccadilly, London.
Scientific American, monthly, Scientific American Inc., 415 Madison Avenue, New York, U.S.A.
The Sky and Telescope, monthly, Sky Publishing Corporation, 49–50–51 Bay State Road, Cambridge, Mass., U.S.A.
Astronomical Calender, Guy Ottwell, Dept. of Physics, Furman University, Greenville, S. Carolina, U.S.A.

Index

Algebraic Logic System, 80
Almanacs,
   Nautical, 12
   Whitaker's, 21
Almucantars, 115, 155
Alt-Azimuth model, 61
   Telescopes, 39, 40
Altitude,
   curves, 115
   importance of, 55
   by pole star, 95
Amplitudes, 51
Analemmatic dial used with Horizontal
     Dial, 196
Arabic numerals, 30
Artificial horizon, 91
Astrolabe, 31, 148
   marking of, 151
   rete, 161
Atmospheric refraction, 88
Astro-Navigation—a basic relation, 47, 82
Astronomical terms, 10, 13
Astronomy,
   in science courses, 6
   branches of, 7
Azimuths, 13, 47, 51, 109, 114, 143

Bubble sextant, 93

Calculators,
   capabilities and accuracy, 5, 95
   growing use of, 2
   misuse of nine figure display, 94
   projects, 4, 5
Celestial latitude and longitude, 228
   equator, 59
Compass correction, 51, 109
   error, 51
Coordinate systems, 59
Course—rhumb line, 99, 101, 103
Cross staff, 34
Culmination, 13

Data for Position Lines, 105

Declination, 10
Departure, 97
Dip, 86

Eccentricity of Earth's orbit, 212
Ecliptic, 14, 59
Ecliptic pole, 228, 230
Ephemerides, 50
Equation of Time, 210
   eccentricity of Earth's orbit, 212
   graph, 217
   obliquity of ecliptic, 214
Equatorial instruments, 61, 63, 64
Estimated position, 83
Errors, effect of, 94
Examples in textbooks, 4
Ex meridian observations, 240
Equinox, 59

Feedback of results, 2, 4
Formulae,
   as calculator programmes, 26
   a summary, 56
   in positional astronomy and in astro-
     navigation, 49
   for spherical triangles, 43
Four Part Formula, 50, 54, 175
   derivation of, 250

Geographical spherical triangle, 78
Great Circle Sailing, 79
Great Cricles on Star Globe, 48

Harvest moon, 242
Hipparchus, 30, 228
Horizon, artificial, 91
Hour Angle, 13, 18, 26
   Local, 16, 54

Index error, 88
Intercept, 84
Interpolation tables, 49

Latitude, 10
   by the Pole Star, 95
   celestial, 228, 230

from a single unknown star, 112
  lines, 10
Local Hour Angles, 16, 54
Longitude, 10
  celestial, 228
  of sun, 216, 218

Mercator Chart, 101
Mercator Star Chart, 137
Meridium altitude, 49, 56, 111, 242
Meridian line, 24
Moon, 12, 71
  eclipse of, 12, 71
  the Harvest, 242
  sights, 90
  extreme declinations, 58, 230
Music of the Spheres, 249

Nadir, 13, 50, 115
Nautical mile, 79
Navigation manuals, 5
Nocturnal, 17
Noon marks, 220
  Polar noon marks, 222
North Pole of the
  ecliptic, 228

PZX triangle, 13, 78, 81
Parallax, 90
Planetaria, 249
Planisphere, 14
  as a nocturnal, 18
Polar graphs, 128
Polaris and the calculator—latitude, 95
Pole Star, 30
  change in position, 228, 230
Positional astronomy, 47
Position, geographical, 78, 81
  from two stars, 82
  line, 83, 104, 105
Postscript, 249
Precession, 57, 228
Prime Vertical Altitude, 54, 56, 110, 117
Projects, 2
Protractor astrolabe, 38, 62
Pythagoras, 30, 44

Quadrant starfinder, 38

Rates of change, 240
Rectangular coordinates, 133–5
Refraction, 70
  atmospheric, 70
  at rising and setting, 73
Rhumb lines, 99
Rhumb line steady course, 101

Right Ascension, 11
Risings and settings, 51, 68
  angles at horizon, 68

Satellites, 247
Semi diameter of sun or moon, 88, 212
Setting circles, 63
Sextant, 41, 42
  corrections to be made, 86
  index error, 88
  latitude by meridian altitude (sun), 49, 56
  low altitude sights unreliable, 88
  model, 41
  practice on land, 91
  telescope and light gathering power, 247
Sidereal, clock, 12, 32, 132
  hour angle, 21
  time, 16, 22, 63, 67
  time reduced to mean time, 22
Sight reduction tables, 47
Spherical triangles, 2, 4, 30, 43, 49, 56, 78, 148, 173
Southern Cross—visible from Europe, 233
Southern hemisphere, 50
Southern latitudes, 135, 257
Spring equinox, 11
Star,
  globe, 10, 47, 48
  maps, 114
  positions in the past, 229
Stars,
  angles with horizon, at rising and setting, 68
  brightest, 256
  when and where to look for, 134, 135, 258
Stellar,
  distances, 254
  magnitudes, 246
Stereographic projection, 30, 148
Stonehenge, 51, 52, 57, 69
Summer Triangle, App. VII, 260
Sun compass, 139, 144, 145
Sundials,
  Various types:
    Altitude dials, 180
    Azimuth dial (Analemmatic dial), 192
    Capuchin dial, or de Saint Regaud, 188
    Disc dial, 186
    Equatorial, 170
    Horizontal, 173
    Pillar dial, 182
    Polar dial, 190
    South facing vertical, 175

South facing vertical-declining, 177
    Sun clock, 170
Sunrise, sunset, 75
Sunshine, amount of, 72
Systems of coordinates, 59, 228, 230

Table of formulae, 49, 56
Tables for curves, 116–27
Telescope, equatorial, 40, 63, 67
    alt-azimuth, 40
Time by the stars, 17
Times of risings and settings, 51
Timing of observations, 90

Traverse and departure, 96
True North by a simple table, 142
Twilight, three grades, 75

Useful information, 252
Useful relations, 49

Vernal equinox, 12

Watch compass, 141, 243
Whitaker's Almanack, 20

Zenith, 13, 81, 114, 115
Zodiacal lines, 200